PENGUIN BOOKS

## *Swimming with orca*

Ingrid Visser was born in New Zealand of Dutch parentage. She has had a life-long love of the sea and the creatures in it. She began researching orca in 1992 and in 2000 was awarded a PhD from the University of Auckland. Dr Visser has compiled the New Zealand Orca Research Identification Catalogue and the Antarctic Killer Whale Identification Catalogue. She has published many articles in both scientific and popular magazines, as well as founding the Adopt an Orca programme and the Orca Research Trust.

# Swimming with orca

### My Life with New Zealand's Killer Whales

Dr Ingrid N. Visser

PENGUIN BOOKS

PENGUIN BOOKS
Published by the Penguin Group
Penguin Group (NZ), cnr Airborne and Rosedale Roads, Albany,
Auckland 1310, New Zealand (a division of Pearson New Zealand Ltd)
Penguin Group (USA) Inc., 375 Hudson Street,
New York, New York 10014, USA
Penguin Group (Canada), 10 Alcorn Avenue, Toronto,
Ontario, Canada M4V 3B2 (a division of Pearson Penguin Canada Inc.)
Penguin Books Ltd, 80 Strand, London, WC2R 0RL, England
Penguin Ireland, 25 St Stephen's Green,
Dublin 2, Ireland (a division of Penguin Books Ltd)
Penguin Group (Australia), 250 Camberwell Road, Camberwell,
Victoria 3124, Australia (a division of Pearson Australia Group Pty Ltd)
Penguin Books India Pvt Ltd, 11, Community Centre,
Panchsheel Park, New Delhi - 110 017, India
Penguin Books (South Africa) (Pty) Ltd, 24 Sturdee Avenue,
Rosebank, Johannesburg 2196, South Africa

Penguin Books Ltd, Registered Offices: 80 Strand, London, WC2R 0RL, England

First published by Penguin Group (NZ), 2005
1 3 5 7 9 10 8 6 4 2

Copyright © text Ingrid N. Visser, 2005
Copyright © photographs remains with individual copyright holders

The right of Ingrid N. Visser to be identified as the author of this work in terms of section 96 of the Copyright Act 1994 is hereby asserted.

Designed by Mary Egan
Typeset by Egan-Reid Ltd
Printed in Australia by McPherson's Printing Group

All rights reserved. Without limiting the rights under copyright reserved above, no part of this publication may be reproduced, stored in or introduced into a retrieval system, or transmitted, in any form or by any means (electronic, mechanical, photocopying, recording or otherwise), without the prior written permission of both the copyright owner and the above publisher of this book.

ISBN 0 14 301983 X
A catalogue record for this book is available
from the National Library of New Zealand.

www.penguin.co.nz

# CONTENTS

| | | |
|---|---|---|
| *Acknowledgements* | | 7 |
| *Prologue* | Ingrid and the Orca | 9 |
| *Chapter One* | A Girl at Sea | 17 |
| *Chapter Two* | Orca Project – First Steps | 31 |
| *Chapter Three* | The Science of Orca | 45 |
| *Chapter Four* | My Doctoral Research | 63 |
| *Chapter Five* | Orca around the World | 85 |
| *Chapter Six* | Dexterous Predators and Dangerous Prey | 103 |
| *Chapter Seven* | Close Encounters of the Orca Kind | 121 |
| *Chapter Eight* | Ben the Wayward Orca | 145 |
| *Chapter Nine* | Filming Orca in the Wild | 163 |
| *Chapter Ten* | The Battle of the Thesis | 177 |
| *Chapter Eleven* | Adopt an Orca | 191 |
| *Chapter Twelve* | Where to Now? | 201 |
| *Appendix One* | Revered and Feared – Orca through the Ages | 215 |
| *Appendix Two* | Extracts from my PhD Thesis | 227 |
| *Glossary* | Orca Terminology | 237 |

# ACKNOWLEDGEMENTS

First I have to acknowledge the amazing orca I have met and got to know over the years. My whole life revolves around them and they have enriched it beyond words.

On the people side of things, how many people actually run Orca Research in New Zealand? Well, just one – me. But then again, working with orca is never a one-woman job. There are always a lot of people involved. To list them all here would require another book! It seems indifferent merely to say 'You know who you are', but how can I thank all the hundreds of people who have sent me orca information, orca photos, or called me to tell me where the orca are? So please forgive me for not listing you all by name. It isn't because I don't care, or don't remember who you are – it just isn't possible to list everyone.

Then there are the research assistants, who have slaved over dull and mundane jobs like data entry; all my sponsors (who are listed on my website), who have made the research project feasible; those folks who have provided me with a place to stay overnight after a long day out on the water; anyone who was involved in the rescue of stranded orca; and scientists who have sent me information from overseas. The list goes on and, of course, includes Geoff Walker and Rebecca Lal at Penguin who have helped me through this process and encouraged me.

I give a special thanks to my dad and step-mum; they continue to support my passion for these animals in too many ways to describe.

Suffice it to say to *each and every one of you*, a most heartfelt thank you. This book is for you.

*Dr Ingrid N. Visser*
www.orcaresearch.org
**0800 SEE ORCA**

# PROLOGUE
## Ingrid and the Orca

I'm not sure which question I get asked more often: 'When did you first become interested in orca?' or 'Are you ever afraid when you swim with them?' As I have loved orca ever since I can remember, I've never been afraid of them. Thinking back, it's hard to say exactly when the fascination became an obsession, but I can remember when I was really young, about six or seven, knowing that one day I wanted to work with whales and dolphins (cetaceans). At that age I was fascinated with all types of animals, but it was cetaceans that really grabbed my attention. I was asked by one of the other kids at school if I wanted to swap a picture of a dolphin, from a magazine, for some of my treasured animal magazines. This little girl turned out to be pretty shrewd, as she managed to achieve a trade-off which saw her obtaining a whole pile of animal magazines, and me with just one dolphin picture. But I was happy, and in fact still have that dolphin picture tucked away somewhere.

However, it was to be many years later, while I was working on my Master's thesis and studying oysters at the Leigh Marine Laboratory about two hours' drive north of Auckland University, before I got my first up-close-and-personal encounter with orca. In the past I'd only seen them at sea while on a yacht, just off in the distance and therefore nothing to write home about – but this time was different. I was staying in a little hut right on the beach edge where it was quite common to see dolphins during the day. I'd been out for a swim with the dolphins before breakfast, and I always kept a mask, snorkel and fins at the ready, just in case.

On the day of this first orca encounter, in May 1991, I was up in the lab when a call came out that there were orca sighted off the beach. I ran as fast as I could down to the hut, grabbed my snorkelling gear and rushed towards the water. All in all it only took me a few minutes to get down to the beach but by this time there were already a fair number of people from the laboratory gathered around. Some of them were clapping their hands together and barking like seals, no doubt inspired by the recent David Attenborough documentary which had shown orca hunting for seals on beaches around Patagonia in South America.

The thought of becoming a snack for a hungry orca didn't deter me for a moment, and I immediately leapt into the water and started swimming. I couldn't see any orca anywhere, and was beginning to wonder if I'd fallen for a prank as I peered in vain through the cloudy water, trying to spot one of them. To counteract the visibility problem I lifted my head out of the water and looked along the surface, and sure enough there was an orca heading straight towards me. I could tell it was an adult male, because as I watched his fin rose higher and higher out of the water until it was towering above me. I quickly ducked my head back under the water, but still couldn't see him. I looked up again for a split second, but his fin had disappeared from view. I dove back down, still hoping to see him, but he was a no-show. I could, however, clearly hear him calling to the other orca. I kept coming up for a breath, and then diving below, grabbing at rocks and weeds on the bottom to stay down longer.

At one stage, as I was heading back up for a breath, I just about had a heart attack, because between me and the surface was an orca. She was on her side looking down at me. I was entranced, but already on my way up, so ended up popping through the surface like a cork, right next to her. She came up for a breath at the same time, and I kept my eye on her. She was swimming past me, in a curve, so I turned with her, and dove down below the surface again. She did the same, and came a little closer. I could see she had something in her mouth, but I wasn't sure what it was – perhaps a sack or something? It was certainly flat-looking, and not fish-shaped. She swam right around me, in a complete circle, with me in the centre, turning to follow her. Then she turned and swam off in the direction the male had been travelling. I thought my experience with these orca was over for the day, but I dove down to hear their haunting calls one more time.

It was suddenly *déjà vu* as I came back up for a breath and there she was again, directly between me and the surface – but this time accompanied by a calf. I couldn't believe my eyes as both turned towards me and surfaced as I did. It was a thrilling moment, to be trusted enough by this female that she had brought her calf over to check me out. And this time it was the calf that had the 'sack' in its mouth, and I could see now that it was a stingray (this was the first hint I'd got that New Zealand orca preyed on stingrays). I was mesmerised by the pair as they swam past me, again in a curve. Once more I turned with them, trying to keep both within my field of view, as I felt that if I broke eye contact the magic moment would be lost and they would turn and leave. Then the calf swam fast in the direction I was surfacing and suddenly it was a game for us both, with the female watching to make sure I didn't make a silly move. The calf was circling around and around me, just below the surface, while I was in the centre, spinning as fast as I could get my arms and fins to rotate me. I was laughing excitedly and it was difficult to believe that this was all actually happening. After what seemed a lifetime (but must have only been a matter of minutes) I was so dizzy I had to sever eye contact with the calf to try and regain my sense of balance, and it was as I had feared – it was all over. The calf stopped in the water, waiting for the female to catch up, and then they both swam close by me on their way out of the little bay and off to continue their lives. I remembered how I had read somewhere that certain American Indians believe that wherever they personally are is the centre of the universe, and for that fabulous moment I fully understood what they meant.

Since that incredible first encounter I have spent over twelve years working with the New Zealand orca, and seen orca in six other locations around the world. I have had some of the most amazing experiences with orca in New Zealand waters, and know many of them by sight – and name. I have witnessed things other people only dream of, and interact with orca on a regular basis. For instance, there is one particular orca, known to me as Ben, who is often intrigued when I am engaged in my underwater research. One time he 'snuck up' and hovered just behind me, out of my line of vision, as I floated in the sea. I was following other orca and observing them while taking notes with waterproof paper and pencil, and composing pictures with an underwater camera. Apparently, according to Ben, I was spending far too much time in one spot, as he seemed to get bored and let out a high-pitched squeal right beside my ear – and of course I nearly jumped right out of my skin, not realising he'd been there all along. Spinning around, I discovered Ben floating right behind me, peeking over my shoulder. Recognising who it was, and what he was doing (just watching me), I turned back to concentrate on the other orca. Well, Ben was having no more of that! He let out a big burst of bubbles and then swam right in front of me – so close I could look him right in the eye and could have reached out to touch him. It was as if he was saying, 'Watch *me*, not *them!*' So I took his photo and spent a bit of time watching him, as he swam around and above the other orca which were hunting below us. At one stage he brushed right past me, and actually rubbed his shoulder on mine, pausing and rolling over to look at me from upside down. It was another magic moment.

Ben and I have known each other for more than ten years and I have complete trust in him, although he is over six metres long, weighs in at over 4500 kilograms and is armed with teeth which could potentially kill me. He certainly seems to recognise me and that may be because I once spent an intense twenty-four hours on a beach helping to rescue him from a stranding – but that part of the story comes later. To begin with, I will tell you a bit about me, my life, how I came to study orca, and how those studies are actually carried out.

## CHAPTER ONE
## A Girl at Sea

I was born in Lower Hutt, New Zealand in February 1966. Eighteen months later my parents had another daughter, my sister Monique. Although both Mum and Dad were born in the Netherlands they had emigrated to New Zealand in 1959, deciding that this was the place they wanted to make their home.

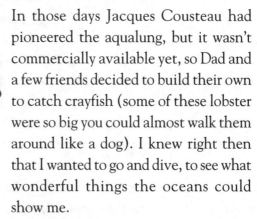

In those days Jacques Cousteau had pioneered the aqualung, but it wasn't commercially available yet, so Dad and a few friends decided to build their own to catch crayfish (some of these lobster were so big you could almost walk them around like a dog). I knew right then that I wanted to go and dive, to see what wonderful things the oceans could show me.

About this time Dad built a small swimming pool for us in the backyard and it became my new universe. Mum gave me a mask and snorkel and I spent many hours in this little box, perfecting diving to the bottom, which was in fact so shallow that I could barely submerge – but I was in heaven.

My sister Monique (left) and I, October 1968. Monique has been out with me to see orca, and I have named an orca after her. **Frits Visser**

When I was seven we shifted to a small 100-hectare farm which was running about 400 sheep and forty head of cattle, located just north of Wellington in an area called Makara. This was an ideal environment for two young children with lots of animals around, and it also wasn't far from the sea. I got my first horse, and even had the chance to take him swimming in the ocean. It was here that my love of animals really blossomed, and it wasn't unusual to see me out and about on the farm somewhere with Dad, collecting frogs from ponds or rescuing birds that had fallen out of nests.

I have often thought that a rural upbringing is an ideal one for a growing child. They not only get to experience the lessons of birth and death early on, but they also learn to have respect for life in general. I remember the time I found a small field mouse that had fallen into a large drum. It was trapped at the bottom and had no way

I was about seven years old when I learned to snorkel. **Frits Visser**

of escaping. I figured all I had to do was put my hand in and it would climb on, and I would be the heroine, having saved its tiny life. Well, the mouse had other ideas about this giant five-fingered monster reaching down to do heaven knows what, and it decided to launch an attack, biting me so hard it drew blood. I belatedly realised that perhaps the mouse would prefer it if I just tipped the drum over for it to run away, rather than be grabbed by my big mitt.

Having proved his ability in running a little farm, Dad now decided to try some real farming and in 1975 bought a larger

240-hectare property where he produced wool (we had 4000 sheep) and beef (135 cows). Situated in the Pohangina Valley, near Palmerston North, the farm was slap-bang in the centre of the North Island and was about as far away as you could get from the sea in New Zealand. I really missed the ocean, and just loved it whenever we could get to the beach.

I was hungry for knowledge in general and consumed books at a very fast rate — in fact, it was quite typical for me to read a book every couple of days. I read about all types of things, and started what I thought was a great book collection, cataloguing it just like a real library. During this time Dad bought me a book called *The Boy Who Sailed Around the World Alone*, by Robin Lee Graham, which would turn out to have many parallels in my own life. It was filled with wonderful illustrations and became almost a bible for me. I read it so many times the spine started to fall off. I knew the route he had taken, and all the characters in this real-life adventure, off by heart. I dreamed of undertaking something just as courageous as this young man. His book still graces my shelves at home, and is occasionally taken down for a loving glance.

At school I quickly realised how very different I was from the other kids. For a start, I came from a European background and therefore had different food in my lunchbox (say, a salami sandwich instead of cheese and marmite); then my name was so very different from their more typical Kiwi ones such as Mark, Chris, Lisa and Jane; and my parents were apparently not 'real' farmers (many of these children came from third- or even fourth-generation farming families). Although I studied hard, got good grades, and liked the learning side of school, the kids picked up on my feelings of isolation and amplified them, and I became an outcast. In reality they were no more cruel than kids today, and I can't say that I was picked on any more than other children, but it was hard and I never felt as though I fitted into any school, or any sort of social group, after that. My early experiences with those children led me to believe that I was a social misfit, and I just came to accept it as a fact.

When I was fifteen my Dad decided it was time to get out of farming, and to head further up the North Island, past Auckland, to Northland. This is the narrowest part of the country, lined with beaches and rocky coastline, and is known as 'the winterless north'. It really doesn't have a true winter, as it is unusual to see a frost, and it certainly doesn't snow. Finding a block of land on the coast near the fishing village of Tutukaka was a dream fulfilled for him when he retired.

We had a 9.7-metre fishing boat called *Pisces* which Dad used for big-game fishing. I often went out with him, and must confess that I once actually caught and killed a beautiful blue shark. Of course, I'm ashamed that I did this now, and can only plead ignorance during that period of my life. Dad doesn't fish for big game any more either, and throughout the last few years he had the boat he only did tag-and-release fishing. However, I made sure I wasn't ignorant about the whales and dolphins that we saw on these trips. I had bought every whale and dolphin book I could find in New Zealand, and was able to identify all the different species from these reference sources. We saw sperm whales, humpback whales, minke whales, Bryde's whales, and both common and bottlenose dolphins, although we never did spot any orca.

Now they had retired, Mum and Dad decided that it would be nice to take my sister and me back to the Netherlands and spend a few weeks looking around. When the travel agent came back to us with a price Dad nearly had a heart attack and said something that changed my life forever: 'I could buy a boat and sail around the world for that much money!' So here began an exciting new episode in my life.

Dad started scouring the country for a suitable boat in which to sail back to Europe. He was on a mission, and *nothing* was going to stop him – but it took him over a year before he finally found the right boat. Called *Wai-O-Tira*, which loosely translated means 'traveller over water', she was a steel ketch (two masts) and seventeen metres in length. She was incredible, built to sustain heavy seas and to travel around the world. Each day Dad and a carpenter worked on all of the internal fittings that had to be completed. Meanwhile, Mum was busy making sail covers, sail bags, cushion covers, squab covers, and all the countless other little things that seemed never-ending on a boat.

When our future home was getting close to completion Dad decided it might be a good idea to learn how to sail! So he got together a group of fellow fishermen, along with Mum and Monique and I, and we set off on what was supposed to be a week-long 'shake-down' cruise to Great Barrier Island. Once we got out of the harbour we tried to put up the sails. Nobody on board had much of an idea about sailing, so there was a lot of yelling of advice and suggestions

Wai-O-Tira, the yacht I sailed around the world on for four and a half years. I travelled with my parents and younger sister, visited 50 countries and sailed 52,000 nautical miles.
**Ingrid Visser**

flying about. Finally, after about an hour, we got some sails up, only to discover that they were on, but upside down. So down they all came, and up they went again, and this time the right way. Our week-long shake-down cruise diminished into a day-long learning process. As it turned out, we never did do another shake-down and ended up leaving for Australia with only one other day's sailing under our belts.

Although we didn't have much practical experience we did consider other critical factors, including navigation, and attended a course on how to navigate with a sextant. It was a difficult course for all of us, but I found it the hardest (not knowing then that I had a form of dyslexia – which in my case means I have trouble comprehending and remembering numbers, often swapping them around without realising that I'm doing so). Dad also decided that we would have a satnav (satellite navigation) on board, a predecessor of the present GPS technology. At the time it was a very sophisticated instrument, giving you accuracy to within four kilometres (these days, even a cheap hand-held GPS will give accuracy to within 100 metres). The satnav was the most expensive piece of equipment on the boat, including the radar and radio.

When it came time to pack our personal gear, even though I knew it wasn't a good idea to take books on a boat (a recipe for instant wet pages), I decided that I couldn't do without a few of my whale and dolphin favourites, and of course Robin Lee Graham's book about sailing around the world. These were carefully packed into plastic bags and stored in the cabin that I was to share with Monique. We had three cupboards, each totalling approximately one metre square, in which to store all our clothes and belongings.

Plenty of food was packed aboard, primarily in cans, but also dried stores and a small amount of frozen goods (we had a little freezer which ran off the engine, although it never did work properly and ended up turning into a storage area for our fresh fruit and vegetables). I had no idea how long we were going to be away, but that was the least of my worries. All I could think of was getting out there and embarking on the type of adventure most people only dream about. As I was only sixteen when we left, I carried on my schooling by correspondence, while receiving lessons which could never have occurred in a classroom.

Finally, on 10 June 1982, we set off. The story of our lives aboard the boat and our experiences sailing around the world would fill at least another book, so suffice it to say here that I got terribly seasick (and still do!) and that the journey was everything I could have dreamed of, and much much more. I dove with sharks, swam with dolphins, had a pet monkey, and met people who had never seen a white person before. We ran the yacht aground, got caught in unimaginable storms, were adrift for days in calm weather, sailed through sandstorms and were chased by pirates. Offers were made to exchange me for camels, Dad got electrocuted and nearly died, I was mugged, and we were nearly run down by a cargo ship. We were boarded by military, chased by warships, deported from one country and invited to live permanently in another. The incredible cetaceans we encountered on our travels were as many and varied as our other exploits.

When we'd eventually completed our circumnavigation and pulled back into Tutukaka harbour in November 1986, the trip had had its full share of trials and tribulations, but if nothing else it taught me to stick with something and to keep going at it until it's finished. People are often commenting on how lucky I am – 'You're so lucky to have sailed around the world' or 'You're so lucky to be studying orca' – and at these times I remember what Dad used to say when we were on the yacht: 'Ingrid, the harder you work the luckier you will get!'

Now that I was back home after four and a half years of travel it was time to decide what I was going to do with my life. I had some idea of what I wanted to do, but back then there was no such thing in New Zealand as a true whale biologist. So my early infatuation

When we were living on the yacht we had a pet monkey called Manis, which means sweetie in Indonesian. Here we are with a cat that she would groom. **Frits Visser**

with studying cetaceans slowly evolved into visions of becoming a veterinarian who would help dolphins in captivity return to the wild, and be there on the beach when whales and dolphins stranded. Little did I know then that I would get my chance to help with both causes, but not quite in the ways I had anticipated.

I set about fulfilling my ambition of becoming a vet by attending the only vet school existing in New Zealand, at Massey University in Palmerston North. While living on the yacht I had learned to become independent and self-disciplined, as well as secure in the knowledge that I was a capable person – but then who wouldn't, when at the age of sixteen you had learned to sail a yacht around the world? However, once I was at university I discovered that none of that self-belief made any difference. At the very beginning of our first year we were told to look at the person sitting either side of us, as the chances were that either you, or they, wouldn't be there next year, because either they would fail, or you would.

Needless to say I found it a demoralising experience, and I ended up thinking that I wasn't going to make it to second year, which actually proved to be the case – my grades weren't good enough. Vet students were expected to get at least A grades, and if you got a C, or heaven forbid a D, as I had, then you didn't have a chance to continue. So I drifted into that zone where many rejected New Zealand vet students go – zoology.

Ironically, failing to carry on with vet school turned out to be the best thing that could have happened for my future career, and I now consider that failure a blessing rather than a curse. Suddenly I was able to learn about the environment and animal behaviour, instead of being forced to do experiments on live animals. However, although I really enjoyed my zoology studies, I was frustrated that Massey didn't offer any papers which were strongly biased towards the subjects of marine biology. My time wasn't wasted though, and I learned a lot and was influenced by many of the lecturers and staff. In particular, our animal behaviourist Dr Claire Veltman, one of the few female academics to teach our classes, had a profound effect on my way of looking at ethology (the study of animal behaviour). She taught us to look at things for ourselves, to question what we were taught, and to think outside the square. She really inspired me and motivated me to consider the possibility of studying and working just on animal behaviour, which was the core of my interest in animals. One of our guest lecturers, Professor John Craig, who talked to us about pukeko, a native New Zealand bird also called the purple swamp hen, was to become a strong influence on my subsequent academic life as well.

By this stage I had discovered that I was dyslexic, and this knowledge helped me understand why I'd experienced so much trouble coping with first-year vet – the numbers had simply been too much for me. In fact, my dyslexia only seemed to grow worse, as it took me more than a year to learn my new phone number. When people asked for my number I would lie and say that we had recently changed it, which was why I didn't know it. These days I just admit that I can't 'do' numbers, although I was comforted to read recently that dyslexics are over-represented in the top ranks of artists, scientists and business executives, and that they are often skilled problem solvers, coming at solutions from novel or surprising angles and making unorthodox conceptual leaps.

Due to the lack of marine biology papers at Massey I had decided that I would transfer to Auckland to finish my undergraduate zoology degree. It would still be a Massey University degree, but would also comprise University of Auckland papers in marine biology. I chose Auckland because it was closer to Tutukaka (only three hours' drive), meaning I would be able to go home on weekends.

While living in Auckland I had the opportunity to attend the launch of Sir David Attenborough's book, *Trials of Life*. What's more, I actually got the chance to talk to him and tell him how I'd been intending to go and study the orca featured in his recent documentary – the ones which were taking seals off the beaches off Patagonia – although I thought studying there might be quite difficult now, with all the media exposure that those orca had received. He agreed and also told me how remote the location was, and that other logistical problems would severely limit my chances of success. Why didn't I work on the New Zealand orca? he asked. I was stumped – it just hadn't occurred to me that this could be a possibility as I had always believed I would have to go somewhere overseas where there were already established projects, rather than try to set up my own.

I spent the rest of the evening mulling this over. Why couldn't I set up my own project? I knew I was capable of it. The only thing that was stopping me was funding and, of course, finding an institute that I could work with, a university that would be willing to help a student with such a project. It turned out that the funding side of things was going to be hard work, while the university side started out as a nightmare, but by the end of that night my course was clear.

After completing my combined Massey–Auckland BSc degree I transferred completely to Auckland to do my Master of Science. I took papers that built on the few marine biology papers I had done for my Bachelor's degree and those that would give me a stronger background in marine biology and in conservation issues. I met up with Professor John Craig again during this time. He was to become instrumental in my study of orca.

For my Master's thesis I studied the growth rates of commercial oysters, which gave me a basic understanding of how to design experiments. I examined the effects of density of oyster numbers, and their relation to the tidal height. I spent a year working on an oyster farm, owned by Jon Nicholson and Jim Dollimore, who provided me with all the oysters I could ever need, and also offered me valuable assistance out in the field. They told me that they occasionally sighted orca out in the harbour, when they were travelling to or from the farm. I was never that lucky myself, but in years to come I would discover why orca were coming into the harbour, and what time of the year would be best to find them there. In the meantime I worked hard at getting my orca research project under way. I had heard rumours about orca sightings around the New Zealand coastline, but this was all the information I had to work from at that very early stage. It was during this time that I had my first memorable encounter with orca.

# CHAPTER TWO
## Orca Project – First Steps

After my thrilling encounter with the female orca and her calf off the beach at the Leigh Marine Laboratory I was on a natural high for weeks, and nothing anyone could have said or done could have brought me down. I was telling everyone who crossed my path about my wonderful experience, and some I told so many times they must have been heartily sick of the story. About this time, while still working on my Master's, and pepped up by the contact with the orca, I became extra serious about seeking funding for my orca project. I went to see a few people, including Professor John Craig, to ask their advice about how to get things started. John suggested that I contact other scientists working with cetaceans and ask them how they had got their particular projects launched and what I would need to do to get things rolling here.

In the meantime, while still struggling hard to get my thesis written, I used pictures of orca I'd put up above the computers in the lab as motivation. Each time I was frustrated or depressed over how long the writing-up process was taking, or how I really didn't have a clue as to what the results meant, I would look up and think, 'From oysters to orca', to spur me on.

Only a matter of weeks after I had written to a number of leading cetacean scientists I got a letter back from a researcher. He was very encouraging and suggested I contact Bill Rossiter, who was with an organisation in the US called Cetacean Society International (CSI). This was the second time Bill's name had come up as a potential contact in less than a week. A coincidence? Perhaps, but one that probably shouldn't be ignored. I sat down and wrote to Bill that very day, explaining how I was hoping to set up a study of the orca off the New Zealand coast, following the lines of a Pacific Northwest orca study, and that I was seeking funding for a camera and lens. Looking back now on the letter I sent, I see how unprofessional it was and how much an amateur I was at begging for funding.

Still, perhaps Bill saw through my naivety, and realised that I was determined to get this study up and running. I received a reply from him – again, only weeks after I had written (as this happened long before email, to get a reply within a month was a thrill, but within weeks was nothing short of a miracle) – and contained within the envelope were some pages ripped out of a camera magazine. Bill had highlighted some cameras for sale, saying that if any of these were suitable CSI could help out with purchasing a body (i.e. without the lenses) in the US, and shipping it out to me in New Zealand. To this day I still don't know if it was Bill's personal money that started it all, or if it indeed came from the coffers of CSI, but either way I am eternally grateful to Bill, and his wife Mia, who have continued to show their support for my research project since that very first contact.

Suddenly I knew I was on a roll, so I splashed out and bought a personalised vehicle number plate. They had just become available in New Zealand, and I was keen to get one with ORCA on it, so that *when* (not *if*) I obtained sponsorship for a vehicle I'd be able to use this as a type of advertisement. I wasn't quick enough to get ORCA (I never did find out who got that one), so I bought ØRCA instead, which looked just fine. I pinned the plate on the wall of my hut, where it drew the attention of all my visitors. When I'd tell them it was for the sponsored four-wheel-drive vehicle I was going to get one day, I'm sure most of them thought I had my head in the clouds – but hey, there's nothing wrong with setting your sights high!

It was also around the time I was working on my Master's thesis that I became strongly involved in the whale rescue group Project Jonah and their ground-breaking work in the re-floating of stranded whales. Through them I was invited to attend a conference about marine mammals held in Christchurch by the Department of Conservation (DOC), the branch of government in New Zealand officially in charge of marine mammals – both those out at sea and those stranded on beaches. Legally, DOC had full jurisdiction over strandings, and it was regarding this aspect of the meeting that we were invited to contribute. They were going to discuss methods for killing whales and dolphins that stranded, and had asked Project Jonah to present a talk on saving them. It was here that I first met Richard Oliver, then a senior skipper and the sea operations manager for Kaikoura Whale Watch, a company operating from the tiny village of Kaikoura on the upper east coast of the South Island.

Richard extended an invitation to the Project Jonah members to come to Kaikoura and check out the whale-watching. He was familiar with the Pacific Northwest orca studies so offered some helpful suggestions regarding my plans to set up a research project to study New Zealand orca. From him I learned that orca came through Kaikoura waters mostly in the summer months, and he said that if I was interested he could probably arrange a seat for me on some of the boats going out to watch whales. It was the perfect chance for me to get some sea-time, and on boats that were dedicated to looking for cetaceans. I decided there and then to take up Richard's offer, but first I would have to go back up north and get my thesis finished.

It seemed that everything was happening at once, because soon after I got back to the lab I received a fax from a man I had met at another conference, called Rodney Russ. He ran eco-tours to the Subantarctic Islands and had spoken about his support for research conducted in this area. I had approached him and asked if he'd ever seen orca off the islands, but his reply was disappointing. He'd only heard of them twice in three years, but I offered to crew on one of his boats, should he ever need staff, in return for the chance to look for orca. Rodney had a boat leaving soon, for six weeks, and he wanted to know if I was interested in coming along. I wouldn't get paid, of course, but I'd get the opportunity to look around that I'd been after.

Now I had to make a big decision – was I going to put my Master's thesis at risk? Would I get it finished before the university deadline? And of course I still wanted to go to Kaikoura and take up Richard's offer as well. Was I going to run the further risk that if I went on this subantarctic tour I might not even have time to go to Kaikoura this season? You bet!

I went to my supervisor, and he recommended that I didn't go. He thought that I was taking on too much (and boy, was he right!) – but I'm known for my stubbornness, so I dug my toes in (then my heels, then my feet, and then my whole body) and said that if he didn't let me go I would quit. If I didn't take this chance to go to the Subantarctic it was unlikely that I would get another one. He told me not to be so silly, that I only had three months of work left, and then I could go wherever I wanted; but I insisted. My parting words as I walked out of his office were that I was about to go and write a letter to the registry informing them that I was pulling out of my studies.

About an hour later, as I was busy packing up my office in a deadly serious mood, he came to see me and suggested that if I was that determined to go on this trip, I should take some work with me and work hard on it if there was any spare time. I agreed, and also promised him that when I came back I would complete my thesis before the deadline. I don't think he believed me, but at least he gave me enough credit to let me try.

Well, in the end I made it back from the deep south in one piece, having been very seasick (I was still affected by it, even after having taken a break from living on a boat), but unfortunately without having seen any orca. But I was back in time to set to work on my thesis again. Every day I put in some very long hours, including Christmas and New Year's Eve holidays – I remember hearing the firecrackers and sirens going off while I was sitting typing away. It was frustrating, but I was determined to get this thesis finished, and not only on time but early, so that I could head south to start my first season of orca research. As it was, I knew I had missed half the season already (orca, I'd been told by Richard Oliver, were most often seen from November to March in the Kaikoura area, and it was already the first week of January), but I was still determined to go. I handed in my thesis before it was due and was on my way.

I spent two months down in Kaikoura, going out on the water whenever there was a spare seat on the boat. Even though I spent many hours watching from the clifftops and going out to sea, I only had one fleeting glimpse of orca. However, it was a great summer and although I was sorry to see it end, I was keen to get back up north to enrol for my doctorate.

You would think I would have learned by now, but I arrived back at university still incredibly naive about how the tertiary education system works in this country. As I was paying to receive certain services from them, I had thought the university was actually going to assist me. How wrong I was. It seems to me that the system is against those who want to do something new, or at least the head of our department at the time was. Marine biology fell under the jurisdiction of the School of Biological Sciences (SBS), which about a year earlier had appointed a new head. The only problem was that he was a microbiologist – and appeared to think that all animals should (and would) be studied from a molecular point of view, and that no other approach was acceptable. Anyone that worked with whole animals, especially those that wanted to study animal behaviour, which wasn't even a 'real part' of an animal, was beneath his consideration. Hardly the ideal environment in which to launch a PhD on the behaviour of New Zealand orca. I may as well have given up there and then – and perhaps if I'd known the half of what was going to be thrown at me I might have.

For starters, I was told that we didn't have a 'whale person' at the university, so I couldn't undertake a study without such a person supervising me. I argued that the chances were the very first person to study giraffes never had a supervisor who was a 'giraffe person', nor the very first person to study cell biology had a supervisor who was a 'cell biology person', and that all I needed – if I might be so bold – was a person who studied *animal behaviour*. Well, as far as the head of SBS was concerned, that wasn't possible either. Besides which, I was also informed that I would need to do a pilot study for a year to show that what I was proposing was a suitable topic – there was no way that *this* department was going to take on a project that might not yield hard results, and if I didn't like it then I might have to consider studying at another institution.

So now I had to go off and study orca for a year – which is what I desperately wanted to do, of course – but I wasn't going to be doing it with the backing of an institution. Again, I was naive enough to think that this wouldn't make a difference. However, the problem I have consistently found with academic institutions is that they place way too much value on themselves and their products. For instance, the public often gets told that observations made by non-scientists usually have little or no value. This is an extremely pompous attitude. Why would someone who has accumulated years observing the ocean, such as a fisherman, not know if an orca was eating a seal or a salmon, whereas I who had only been trained in a university classroom *would* know the difference?

Needless to say, I embarked on my pilot study already furious with the system. Originally, I had wanted to carry out this study as part of a Master's thesis but had been turned down, as I was told that it wasn't suitable for a one-year project. Yet now I was being told to go away and do a one-year project so that I could proceed with my doctorate. Something definitely stank, but I was just so focused on working with orca at whatever cost that I was determined to climb over every obstacle in my way.

So there I was, unofficially starting up my research project. It was 1992, and although I had been working part time on setting up the project while I'd been working on my Master's thesis, I was now into it 100 per cent. I was writing articles for magazines, and asking anybody who saw orca to contact me. I also started looking through newspaper archives (discovering that the most productive area to look under was the fish section) and actually found a few articles about orca and other whales and dolphins. I searched through all the major, and many of the minor, papers from around the country, concentrating on those from coastal towns and villages. Wherever I could find a space, I stuck up mini-posters asking after any information about these animals. I put out the word that I was after *any* photographs of orca (as I knew from the Pacific Northwest studies they were long-lived creatures), and I figured that any historical photos might give me some clues about individuals. In addition, I started giving talks to yacht clubs, dive clubs, schools, and so on – basically, anywhere and to anyone that would listen.

I also let it be known that I was interested in hearing about any sightings from the past, especially any that might have dates and locations. I was working on the assumption that New Zealand orca might be predictable, just like orca in other areas. They probably had certain spots that they visited at different times of the year, in search of either food or mates. In this way I hoped to build up a database that would allow me to work out any patterns, which in theory would allow me to place myself in their path and furnish me with the best chance to get out there and do some 'real' science.

It only took about six months before I started coming up with some very interesting results. I began getting records of orca off both the North and South islands of New Zealand. Sure enough, there did look to be a pattern: the orca appeared to be seen more often in the southern waters during the New Zealand summer and the northern waters in our winter, which made sense if they were following warmer water – but *was* that what they were doing? And I still didn't know if the animals that I was hearing about off the North Island were the same ones I was hearing about off the South. Perhaps there were two different groups that happened to come to the coast at different times of the year, and at different locations? This was possible, as the Pacific Northwest researchers had discovered two different types of orca, with quite different lifestyles and different eating habits.

I was starting out cold with this project, without any firm idea of what was really going on. I had heard from various 'experts' (and there were plenty of those around) that there were no orca around New Zealand; that there were only about twenty that travelled around the whole coast; that there were hundreds if not thousands around the coast; and that orca seen around New Zealand waters only came here once before heading off to Antarctica. If you think of a story that could be told about orca, I heard it. But there were no hard facts at that time, and I knew that without hard facts it would be difficult to prove anything, and I certainly wouldn't be able to register for a PhD without some solid results.

Then, bingo! Through Steve Whitehouse from Project Jonah, who had taken a photo of an orca in Auckland harbour in 1987, I got a match with a photograph of an orca from the South Island, off

Kaikoura. This was a major breakthrough. It mightn't seem like much, but to be able to positively show that at least one orca was moving between the North and South islands was the basis for a start. The reason I was so sure that it was the same animal was that he had a bent dorsal fin, giving it the appearance of being twisted. The fin also had a notch out of the front. The combination of these factors made him unmistakeable, earning him the nickname Corkscrew from observers.

Then, only three weeks after that first photo match, I got a call that there were orca sighted in Auckland harbour. I was actually in the city that day, and so was Steve. He arranged for us to go out with a friend of his, Rosco, who had an inflatable boat. I had my new camera from CSI, as well as a brand-new zoom lens bought with money from a grant provided by Project Jonah, so I was all set to go. Off we raced, down to the water's edge, where Rosco was waiting for us. We leapt into the boat and headed off. I was a bundle of nerves – would we be able to find the orca, and would we be able to get close enough to actually photograph them? As luck would have it, they were not far away from the boat ramp, and even though the light wasn't so good (it was late in the afternoon in the middle of winter), we stood a chance at least. When we first located them I was covered in goosebumps – I actually recognised one of the animals.

This was incredible – she had been photographed by Steve back in 1987, when he'd also photographed Corkscrew, and not more than a mile from where we were currently watching them. It was a great thrill, not only to be with the animals, but also to recognise one of them and know even the smallest bit about her life. I asked Steve to name her, and as she had a large nick out of the trailing edge of her fin, he called her Nicky. I finally had incontrovertible evidence that orca were travelling between the North and South islands, and that they were returning to the same locations after a number of years. I still occasionally see Nicky and think back to

Nicky is a female orca so called because of the large nick out of the trailing edge of her dorsal fin. I photographed her in Auckland in 1993 and matched her to an earlier photo (top) taken by Steve Whitehouse in 1985 at the same location. This was my first hint that the orca were returning to the same locations over time. **Ingrid Visser**

that first historic day out. The same week I had yet another breakthrough – a female orca with the top of her fin missing was matched between the North and South islands as well. Suddenly it didn't look as if Corkscrew was an oddball by nature as well as looks, because the second and third matches suggested that there might also be a pattern to their movements.

I felt I now had enough evidence to reapproach the SBS and see if they would accept my project. I was over-optimistic. They still wanted more from me, saying that I hadn't spent enough time on the water with the animals, so how did I know that I could actually study them. It seemed a never-ending vicious cycle – I'd been told that I couldn't enrol while conducting my pilot project, and therefore wasn't eligible to use the university boats (and I didn't have the money to go out and buy one for myself), so I was limited in what I could accomplish out on the water.

I suggested a compromise. I would find a temporary supervisor, which the head of SBS could then approve, and if after a year of being enrolled I had not come up with enough hard data to show that I could in fact study the animals out on the water, I would not hassle him any further and would transfer to another university. He took up the offer, and then told me that he had actually recently appointed a new scientist from America, who was a whale biologist and who would be arriving at the beginning of the following year. If I could prove the study would work, then I could transfer across and continue my study under him. I couldn't believe my luck – until the Head informed me that the new scientist was a molecular biologist, specialising in the DNA of whales. This was extremely disappointing, but I figured he would at least be sympathetic to my study. In any event, I jumped at the chance to finally start my doctorate in a legitimate way, and set about seeking an interim supervisor.

In my mind there was only one suitable candidate – Professor John Craig. I approached him immediately and showed him what I had already done. He warned me that I was taking on a huge project (if only he and I had known *how* huge!), but if I was willing to give it a go he was prepared to be a temporary supervisor until the new guy arrived from the States. I was thrilled, and told him so, and also promised that I would try to be as little trouble as possible for him. I had no idea that there was further trouble to come, and that a fair amount of it would pass John's way, via me. But at the time all I saw was that I had finally been given the chance to launch my project, and I set about officially enrolling.

# CHAPTER THREE
## The Science of Orca

Considering that cetaceans don't sleep in trees or caves and don't build nests, where do you start to look for them? They don't leave a set of tracks, or stick to a defined trail or path, so how do you actually follow them? As most cetaceans spend less than ten per cent of their lives at the surface, how do you know what they are doing the rest of the time? Although the larger cetaceans produce faeces the size of an elephant's (or bigger!), you don't find it sitting there in a pile (in fact it doesn't even float), so how are you going to discover what they eat?

Although some techniques from land-based studies of animals can be applied, they often need to be highly modified and adapted to suit working in the marine environment. Originally though, scientists started working in two areas where their standard terrestrial methods could be easily used: at strandings and on the whaling grounds.

Humans have been finding stranded cetaceans on beaches since we first came to live on the coast. They were a bonus to our diet and for many peoples were also considered a gift from the sea. The dead animals provided much-needed, and easy-to-gather, protein. Fat could be used for curing skins, as fuel for lighting and for preserving food. The huge bones of some species were incorporated in housing structures, while the smaller bones were modified and utilised for various items such as needles; the teeth were highly prized for ornamental reasons. Some peoples camped beside a stranded whale until they had completely consumed every part.

Scientifically, strandings provided some of the first instances for examining these large animals up close, and so enabled them to be investigated beyond a superficial level. Beachings attracted the attention of the general public as well: one that occurred on the western seaboard of Europe in the 1800s became the centrepiece for a circus which included giraffes and elephants – then exceptionally rare in the area. Strandings continue to provide us with valuable information about cetaceans, and in some cases our only knowledge of certain species comes from stranding material. Two of these are beaked whales (with a long snout, somewhat like a dolphin): Perrin's beaked whale, *Mesoplodon perrini*, which has recently been described based on findings from only five strandings, and *Mesoplodon traversii*, which is only known from three partial skulls and is so rare it doesn't even have a common name.

Carcasses can provide us with basic information like the internal structure of these animals – including skeletal features – and in cases where the animal is pregnant, give us an insight into their reproduction. A wide range of age groups strand, or are washed up dead on beaches, and this can present us with a broad cross-section of the population. Cetaceans can strand as singles, pairs (typically mother–calf pairings), or in large groups called mass strandings.

In the case of orca, one of the earliest recorded strandings is from 1783 on the coast of the Netherlands; however, sub-fossil evidence shows that orca strandings have been occurring long before that, at least pre-sixteenth century. In New Zealand, although exact records weren't kept in the early days, we know that orca were stranding here prior to 1880 and have continued to do so. From my investigations I have found that New Zealand has one of the highest stranding rates for orca in the world.

Once the whaling industry took off, and particularly once it started to collapse because the number of whales killed was outstripping the number of animals breeding, scientists got involved. They wanted to find out why there weren't so many whales around. This seems pretty obvious to us now, but back then whales were so numerous that it just didn't seem possible that humans could decimate the populations. Scientists were also interested in how often they were breeding, at what time of the year, and additional information such as what the whales were feeding on. The whaling industry gave them the perfect opportunity for studying this – multitudes of dead carcasses, and all getting chopped into bits right in front of their eyes – all they had to do was start recording the data. And record they did. To give you an idea of the numbers, estimates place the number of sperm whales taken just in the five years between 1957 and 1961 to be in excess of *21,000* animals. Of course, other species of whales were taken as well, including, but not limited to, fin, sei, southern right, humpback and blue whales.

In many cases orca escaped the lethal arm of whaling. Overall they were just too small to bother with; although they definitely *were* taken – for target practice, or when there weren't other whales around, or if they were attempting to feed off other whales which had already been slaughtered. However, some orca-specific whaling did occur – particularly in Antarctica. Russians hunted for orca during two Antarctic seasons (1961–62 and 1978–79) and took 323 orca. The Antarctic whaling season occurs only during the Antarctic summer, from November to February, so is in fact only a few months long. In the Antarctic season of 1979–80 other Russians took a further 906 animals, giving a total of 1644 orca officially taken. We have no idea how many orca the Japanese or other nationalities have taken; in the early days they were not required to report takes of smaller cetaceans and we now also know that many reports were fudged – with false numbers given to conceal exceeded quotas.

Catching orca for scientific purposes, the Russians took body measurements, noted physical characteristics and the stomach contents of these animals, and both groups of whaling scientists proposed some radical new ideas. These included two new species of orca – *Orcinus nanus* and *Orcinus glacialis*. *Orcinus nanus* was described as a 'dwarf' orca, based on the animals maturing at a much smaller size than 'typical' orca. *Orcinus glacialis* was also described as smaller than typical orca but it lived close to the ice edge – hence its name.

Although whaling continues to be conducted today by some countries (Iceland, Norway, Japan and Russia as I write this), both their science and published findings are weak to say the least. Commercial whaling is officially banned by the International Whaling Commission (IWC), although this same organisation has issued 'scientific whaling' permits, and whaling continues under that guise. The theory is that the animals harvested are investigated scientifically – and the theme of the 'science' is to look at the whales to see if they have reached levels suitable for harvest. Therefore it is a self-sustaining industry. Also, the whalers claim that this science is very expensive, so they sell the by-products (whale meat) – and reap huge profits. Although only certain species of whales are allowed to be taken for scientific whaling, basically all species known to be in the waters where the whalers are working are turning up in their nations' supermarkets (including fully protected and endangered ones). Of course, this is a very simplified version of the whole scene, but it gives you an idea of what is going on out there, which is that the days of whaling are far from over.

By the late 1960s and early 1970s scientists were starting to work with live cetaceans – and not just those in captivity, although there have been integral links between all methods of learning about cetaceans. For instance, the first-ever dolphin kept in captivity was an animal that had stranded alive. It was transported in a glass tank on the back of a horse cart, as part of a travelling sideshow, and when it died doctors investigated its carcass. But now it was the era of fieldwork – studying cetaceans in their natural habitat.

This type of research was based on a novel concept first offered scientifically by biologist David Caldwell, who saw a dolphin with the top of its fin missing in the waters off Florida in the mid-1950s. He re-sighted it over a four-month period and published a short article in a scientific magazine in which he stated that although he had no proof this was the same dolphin, he felt 'confident that it was the same animal' and that 'there seems little likelihood that such an extensive and characteristic disfigurement would be duplicated coincidentally'. This was followed by another similar-style article in 1960 in which the subject was a humpback whale. The researchers also suggested that an individual cetacean could be recognised repeatedly and over time and distance.

In 1970 scientists off the coast of British Columbia and Washington State (the BC/WS coast) set out to discover how many orca were living along the coast. From the mid-1960s there had been a rapid growth in the taking of these animals for the captive industry (255 orca during this period). Locals and governments alike were concerned that the removals would be unsustainable, so they wanted to know some basic details about the orca – how many there were and where they were going (e.g. were the orca seen off the Canadian coast the same as those seen off the American coast?). The head of the project was Dr Michael Bigg and he set in motion a study which continues today. Over a three-year period a questionnaire was sent to anyone who lived or worked along the coastline the scientists wanted to survey. From the response it was estimated that there were between 200 and 350 orca living along the BC/WS coast – not many when compared with the numbers that had been *thought* to be living there.

To get a better handle on the animals themselves, in 1972 Mike Bigg and some of his colleagues headed to an area which the questionnaires had highlighted as a prime spot to find orca. Sure enough, they found them, and so reliably they encountered them every day of their survey. They took photos while watching the animals and quickly discovered several individuals with distinctive marks, ranging from scars such as bits missing from the dorsal fin to dorsal fins with characteristic shapes such as a bent-over tip – these were natural 'tags' which made it possible to study individuals and learn about their daily lives on a personal level. Taking a gamble, they returned in 1973, at the same time of year, to the same location, hoping to locate these 'tagged' whales. Locate them they did, and so started the first-ever photo-identification (photo-ID) study of orca – a method that has now become standard practice for field studies of a wide variety of cetaceans.

The level to which the individuals are identified is much more detailed nowadays – it is not just the very obvious marks such as nicks, cuts and gouges which are used, but tiny features such as a small white spot below an eye, a pair of black parallel marks across a light grey area, or even just the way an area of pigmentation fades from dark to light. All these characteristics add up to make a type of mug-shot profile for the animal which is then given a code number. For instance, the New Zealand orca catalogue is filed NZ1, NZ2, NZ3, and so on. In other catalogues the animals might have codes which designate which group they come from – in Mike Bigg's study the animals were coded according to the order in which the groups were encountered, so there is K group and L group, and the animals within the groups are then numbered consecutively as they are identified.

Matching animals to their photo-ID can be a very laborious process, but then again it is a lot of fun. Searching for tiny identifying

This is the desk I work at, surrounded by orca paraphernalia I have collected over the years. I spend as much time behind a computer, working on a variety of things like funding applications, data entry, writing scientific and general articles, and processing photos, as I spend in the field working with the orca. **Angie Belcher**

features helps you really get to know the individuals. After a while you start to recognise some of the animals instantly and know exactly where to go in the catalogue to find them. If you make a match then you have some significant new data for your research. You might not have photographed that animal for six weeks, or you might have seen it in a new location. Perhaps it was with an orca you have never seen it travelling with before. There is any amount of information that a basic piece of identification can give you, especially if you accumulate enough of it, and over a long period of time.

If you don't find a match, then you can almost surely assume that you have a new animal for the catalogue. However, there are several factors to take into account. For instance, you must be sure that you have a very good photograph of the individual. For this reason, photo-ID images are 'graded'. A Grade 5 is a clear, sharp photo in which fifty per cent or more of the frame shows the animal and is taken side-on, that is, ninety degrees to the animal and not from slightly to the front or the back. A Grade 2 is an image which is out of focus, at the wrong angle, or in which the animal might just be a tiny speck in the far distance. A Grade 1 is a sighting in which no photograph was obtained; for example, you might have taken photographs but there was a problem in the development.

I use a system which also takes into account the information accompanying the photograph – such as the date and location. Because some of my data comes from the public, via sightings and photographs of the animals, I need to be sure the quality of the information used in the later analysis is robust. For instance, I further grade the incoming data by source as well as the quality of the photograph. A Grade 5 source might be someone who is a whale-watching boat skipper – they can identify an orca, know the sort of data I require, and record the precise time the animals were seen and the exact location, perhaps with a latitude and longitude reference. A Grade 1 source might be a member of the public who doesn't really know what they were looking at or maybe can't really remember when the sighting occurred.

Once you have graded a photo-ID you still need to be sure that your evaluation of the photograph with reference to the catalogue is made in the most error-free way. For example, what about animals that appear so incredibly alike that you list them as the same orca? A 'mis-match' like this results in an underestimation of how many orca are in the population. There are some statistical analyses which can be done to help identify where such errors lie. Based on a kind of photographic 'tag and release' or 'capture and release' principle, you should be able to work out how many animals there are based on how many 'tagged' and how many 'new' animals you photograph.

When it comes to the way you set up and organise your photo-ID catalogue you also need to decide just which part of the animal you are going to use as an identifier. Given that Mike Bigg's project started with dorsal fins and progressed to include the grey area behind the dorsal fin, called the 'saddle patch', this has pretty much become the standard for orca catalogues. Other cetacean researchers use completely different parts of the animal's body – for instance, humpback and sperm whale researchers use the trailing edge of the tail, which they photograph as the animal raises it to dive. In the case of humpback whales the researchers also use the underside of the tail as this often has very distinctive pigmentation patterns. Southern right whale researchers photograph the distinctive 'callosities' (wart-like growths) on the heads of the animals.

Whatever feature the researchers use, they have to be sure that the marks are long-term and distinctive. When it comes to using marks other than the shape of the dorsal fin or the tail – for example, pigmentation patterns or gouges caused by something which has scarred the animal such as brushing up against a rock – you need to be sure they are not missed if you are comparing different sides of the animal. Pigmentation patterns on either side of an orca are not necessarily the same; for instance, an orca might have a white spot below its eye, but only on the left side. So do you only photograph the animals from one side, and consistently stick to photographing that same side, or do you compile photographs of both sides to avoid the possibility of error?

There are proponents and opponents of both methods, and likewise there are good and bad reasons for each. An argument for the 'one-side catalogue' is that it creates a smaller catalogue, and additionally you can be sure you don't catalogue an individual twice – once for its left side and once for its right – without realising you've double-listed it. Furthermore, if you were driving your own research boat (in which the wheel is typically on the right-hand side) you would have the animals on the right side of the boat, and this would result in you taking photographs of their left side as you travel with them. This is precisely the reason why Mike Bigg's group started their catalogue with left-side photographs, and continues to do so. However, a negative aspect of the one-side catalogue is that if you are sent photographs of an animal from the opposite side of what is in your files, you won't be able to make a match. On the other hand, if you use the both-sides catalogue you might have a lot more data to sort through, but you stand a higher chance of rematching an individual. And if your boat has a centre console then you can take photographs of the animals from whichever side they come up on. This is the method I use.

Another aspect that needs to be taken into account is the possibility of missed matching when the animal's original marks have been disguised or removed. For instance, you identify an orca with a notch at the very tip of its dorsal fin, but in subsequent sightings the whole top of the dorsal fin is missing, and therefore the distinguishing mark from the tip, and so you fail to recognise it. This is why the use of secondary features, such as a pair of dark parallel marks on the saddle patch, to double-check the identity of the animal is vital to the integrity of your sightings database.

Being able to identify individuals when out in the field allows you to carry out what are termed 'focal animal follows'. These basically consist of the researcher following an individual, or a small group of individuals, for as long as possible. While following you might record a wide range of data. You could be looking at who the individual spends time with, or who it catches food with, or who it shares its food with. You might look at how much time the animal spends sleeping in comparison with the time it spends travelling or hunting.

However, it must also be taken into account that orca (like all species of cetaceans) spend approximately ninety per cent of their lives underwater, which means that in theory you are only going to physically see an orca you are following for approximately six minutes out of every hour – and that is only if you are looking in exactly the right direction when it comes up. Consider also that a single surfacing is usually around a second long and you have to ascertain a lot of data within that second. Are you following the same orca? Did you get a photo-ID shot of the fin, the eye patch, the saddle patch? Is the orca travelling with anyone else? Does it have anything in its mouth? Within that six minutes you might see nothing more than the animal surfacing for a breath, or you might see it sleeping, but then again you might observe travelling or socialising or hunting behaviours. Identifying these behaviours takes a little training and a *lot* of experience and patience.

Sleeping is often recognised by the orca surfacing synchronously with other members of the group, or surfacing very slowly, and after just a few surfacings a rhythm is usually apparent. Occasionally the animals may lie at the surface, just floating there, a behaviour termed 'logging' because they resemble a floating log. Travelling is just that – the animals moving in a constant direction at a steady speed. This might be maintained for just a few minutes or for a few hours. Researchers typically note the speed the animals are travelling at, the direction, and whom they are travelling with and in what order they are surfacing.

Socialising is a very complex group of behaviours and the definitions vary from researcher to researcher, with the observations also dependent upon the population of orca you are watching. For instance, in the waters off BC/WS the orca often breach (jump right out of the water). Breaching might occur ten to fifteen times (or even more) during a research encounter of a few hours, but in New Zealand you might watch orca for ten to fifteen days and *never* witness a breach. Just why the orca are breaching is something – along with many of the other socialising behaviours we observe – for which we don't yet have clear answers. It is thought that breaching, and other socialising behaviours which cause loud noises on the surface and can be heard a long way distant, such as pectoral fin slapping and tail lobbing (where orca slap the top of the water), may be a form of communication.

Hunting behaviour also differs from location to location and between different orca populations, with techniques ranging as widely as the prey types. For instance, off Argentina the orca come up onto beaches to take sealion pups, meaning researchers are better off remaining land-based; whereas off Norway herring schools form in the springtime and researchers have learned to follow the herring and wait for the orca to turn up. In the waters around New Zealand, however, orca often hunt for rays in shallow water, so whenever possible I get into the sea and watch what is going on down below.

In addition to observing and recording all these different aspects of orca behaviour, the basic photo-IDs, the focal animal follows and the behavioural observations are now being supplemented by DNA 'fingerprinting' of individuals so we can deduce other aspects of their life. The samples are collected in a multiplicity of ways, including gathering faeces (from which stomach and intestinal cells can be separated out later) and whale 'dandruff' – small pieces of skin that slough off as the animal swims around, particularly when it jumps and lands with a smack back in the water. Another method is to collect 'scrub' samples with a small, sterile, rough pad on the end of a pole, which harvests tiny amounts of skin for analysis.

One other technique used to collect skin samples which has proved fairly controversial over the years is biopsy-darting, or just biopsying for short. In a scene reminiscent of the early whaling era, a dart with a sharp hollow tip is shot into the side of the whale to collect a sample of skin and blubber. The dart then falls from the side of the whale as it dives and is collected by the scientist. These darts were originally so big that the animals tended to react quite strongly when they were hit. In some instances they thrashed their tails violently at the surface, or dove suddenly and disappeared, or avoided the boats altogether. Nowadays, however, we have learned a lot from these early experiences and have progressed towards collecting much smaller biopsies which – I can verify from my own experiences while working on a project in Iceland in 2000 and 2001

– in many cases don't even elicit a response from the animal.

Although still controversial, and still invasive, the information we are beginning to gather from this DNA sampling, combined with the long-term data from photo-ID and behavioural studies, will keeping expanding our insights into the lives and well-being of these creatures. For example, from the blubber taken with the skin during biopsies we can look for a wide range of poisons and toxins such as PCBs, DDTs, mercury and flame retardants, many of which are fat-soluble and so are not broken down by the animal's liver and kidneys. Some studies have revealed serious levels of bio-accumulated pollutants – i.e. animals at the top of the food web, such as orca, have high levels of these poisons and toxins accumulated from all the creatures below. One recent study, by Peter Ross from Canada, looked at the orca from the BC/WS coast and found that they have such high levels of these pollutants in their bodies that they can now be considered the most contaminated cetaceans in the world.

Other significant areas of research include the acoustics of orca. When orca research first started off BC/WS, Dr John Ford, one of Mike Bigg's colleagues in the study, surmised – correctly – that population analyses could be made based on recording the calls of the orca. Little did he know that his ground-breaking research would provide a key to understanding their complex social structure. We now know that orca produce a wide range of sounds such as clicks, squeals, shrieks and whistles. Furthermore, it turns out that each population of orca around the world has particular calls which are unique to that population, and within each population there are variations – or dialects – which can be used to identify an animal's individual group. John has proposed that these acoustic differences and dialects could be used by the animals to identify membership within a group and might even be used when seeking a mate, possibly to prevent inbreeding. Using a hydrophone to make the recordings, these calls can be played back later on computer to form spectograms, which look something like the picture on the following page.

A spectrogram of New Zealand orca calls – time is across the bottom and pitch up the side. The two sets of long up-curving lines on the left represent two whistles (harmonics are represented by the thinner lines stacked on top of each other). After the whistles the dark 'block' is a series of very closely spaced clicks, which sound like a creaking door, and the upright lines to the right of the block represent wider spaced clicks called echolocation. The length of time for this train of calls is only 4.5 seconds long.

Other researchers interested in what goes on in the underwater lives of orca use a type of tag with an embedded mini data-logger – a Time Depth Recorder – which is typically attached by a suction cup to the orca's side. While attached, the TDR records details of how deep the whale is diving, for how long it dives, and the speed at which it is travelling. If this data is then superimposed over a plot of the ocean floor all sorts of information can be extrapolated. For instance, if an orca is recorded reaching the bottom on every dive and staying down there, perhaps it is feeding there (on rays or flatfish). If it is only diving to the bottom occasionally, perhaps it is trying to establish if there is suitable prey in the area or is looking for suitable hunting habitat. If it is diving only a short distance below the surface and then travelling at that depth, perhaps it is sleeping or just travelling. If it is diving part of the way down and then spending time at that depth, it also might be hunting for fish who live at that depth or prey that swim above it – seeking their silhouette against the light-coloured water surface.

However, one of the biggest difficulties with tags lies in trying to attach them humanely – in terms of both the physical attachment and the emotional trauma for the animal associated with capturing it. To attach a tag other than a suction cup type (deployed either using a crossbow or a long pole, which is used to slap the tag onto the orca's side when it surfaces) requires the tag to be secured properly. This usually involves some sort of surgical procedure such as a local anaesthetic and punching holes into the dorsal fin through which bolts are passed and secured to the device. Such procedures entail risks which consist of, but are not limited to, infections of the holes, skin infections under the tags, and 'migration' of the tag – whereby the tag moves from its original position to a new one, creating rips or tears in the fin through drag as the orca moves through the water. When Keiko, the star of the *Free Willy* movies, had a large-sized tag attached it migrated, resulting in tears in the fin and infection of the holes. He was quickly treated and the holes healed, and the next tag was much smaller and hardly migrated at all during months of attachment.

Other drawbacks of tags can have a wide range of effects on the data. For instance, if a tag is large it might affect the way an orca swims (and therefore affects its ability to catch prey or keep up with its group). The tag might even affect how other orca in the group interact with the tagged orca – as was found in the case of zebra finches when they were tagged with coloured leg bands: certain combinations of colours turned out to be more attractive to the females, so males who had been fitted with these colours were more successful in finding mates. But the converse could also be true and orca might be rejected from their group if the tag was found offensive. Also, orca who have been captured to have a tag attached might not be able to find their group again, so rendering the data collected irrelevant to their normal daily lives.

Advances are continually being made in the area of tagging, and as computer and battery technology advances it shouldn't be too long before there is a tag that is small enough not to affect the hydrodynamics of an orca and which will require minimal attachment devices. At present tags attached to cetaceans do not transmit through water so must send their signals only when at the surface, and there have been suggestions to implant them below the surface of the blubber. However, at this stage the technology is not quite developed enough as the tags would still have to be surgically implanted and are likely to be rejected by the animal's immune system. Additionally, tagging is still horrendously expensive and the recovery rate of deployed tags is minimal, so they have to be considered disposable. Until we can circumvent all these issues, tagging will provide valuable information, but at a much reduced rate than might be possible in the not-too-distant future.

The study of orca is like working on a living jigsaw puzzle: to begin with you have an idea of what you are aiming for, but it isn't until you get a large number of pieces into place that you can really start to see the big picture. And sure enough, this is what is happening with orca research. Mike Bigg's study, now run by his colleagues John Ford, Graeme Ellis and Ken Balcomb, has been going for over thirty years now and has provided some amazing insights into the lives of the BC/WS orca. It has also established a baseline for other orca studies which are starting to spring up in different locations around the world – including my own.

# CHAPTER FOUR
## My Doctoral Research

When I enrolled for my PhD in the early 1990s I had no boat and only a tiny second-hand car which couldn't have pulled the skin off a cup of hot milk. My findings from the pilot project had suggested that there was no single place around New Zealand where orca could consistently be found, as is the case with one or two spots around the world, which makes it possible to study them from land. Nor did there seem to be any particular time of the year to search for them, even if basic trends had started to emerge. So I was faced with having to begin at the beginning, and that would mean spending much more time on the water. There was only going to be one way to achieve this on a regular basis – and that was to get my own boat.

I didn't know any women who owned their own boat, but I wasn't going to let that stop me. I knew exactly the type of craft I wanted: a Naiad – a brand of rigid-hull inflatable designed to give the best of both worlds in terms of the safety and strength of a rigid-hulled boat and the speed and stability of an inflatable. The Naiad also had a deep V-shape to its hull which meant it could cut through waves as well as surface chop. Although a fair bit of spray could get blown inside the boat on a rough day, I could dress in good wet-weather gear, figuring that it was more important to have a safe boat than a perfectly dry one.

In addition, I calculated that there were going to be nine main areas where I would incur setup costs and eight which would involve ongoing costs. Initial outlays included a four-wheel-drive vehicle, a boat (with engines and a trailer), safety equipment (flares, lifejackets, radio), diving equipment, wet-weather gear, extra research equipment (GPS, hydrophone, etc/), a computer, waterproof storage for all the equipment and an underwater housing for the camera. Ongoing expenses comprised fuel for the four-wheel drive and boat, insurance and maintenance for the same as well as all the other equipment, film for the camera (digital cameras weren't around at that stage), development costs for the photo-ID shots, office supplies and postage costs, communication costs, as well as anything else I hadn't thought of – this turned out to be the biggest cost! No matter how I tried to juggle the budget this was not going to be a cheap project and I still had my student loan to pay off.

So I set about applying for funding. I applied for sponsorship to every company I could think of, and it didn't matter if the company's activities were relevant to whale research or not – if I had an address they were fair game. I tried food companies, petrol companies, sporting goods companies, women's products companies, camera manufacturers and the companies which imported camera gear. I tried shoe companies, chewing gum companies and telephone companies. All in all, I must have written to over 500 businesses, and this was all done without my own computer. Instead I had to go into university to write up the application letters and compile CVs and project outlines, while learning how to use computers in the process (these were still very much the domain of the nerds in those days).

Months later I was still receiving rejection letters. The typical response was that the project was a good one, but that the company in question didn't sponsor individual projects, they were more interested in community-based enterprises. Fair enough, but this didn't get my project off the ground!

I was starting to get pretty desperate and wasn't quite sure where to turn to next. I needed a break, and luckily I was invited to join a couple of friends on a day's diving trip to the Poor Knights Islands, about twelve nautical miles off the coast where I live at Tutukaka. The outing was fun – but not only that, it also turned out to be the turning point for my research project as aboard the dive boat was Jerry Carpenter, then head of the New Zealand branch of Jacques Cousteau's organisation US Divers (also known as Aqualung). Jerry and I got talking and he suggested I send him a proposal and drop in to see him the next time I was in Auckland. I didn't have a trip planned, but I told him I would be down the following week. I wasn't going to let anything stand in the way of a possible sponsorship.

It turned out that Jerry was also in charge of the sponsorship side of the company and he set about kitting me out with a full set of dive gear which I still use. Landing such a prestigious endorsement lent the project enormous kudos when it came to applying for other funding, and Jerry also suggested I get in touch with Ken Goody at Kodak, so my next proposal was sent off to him. Perhaps it was the US Divers/Aqualung logo at the bottom of the application letter or perhaps Jerry had put in a good word for me, but Ken said that Kodak would also support the project. Suddenly I was on a roll!

It was time to start thinking big so I decided to apply to some of New Zealand's major funding agencies. One grant in particular I applied for through the University of Auckland because the application forms didn't permit individuals to apply. So I filled out all the appropriate paperwork, which took weeks of effort to obtain the necessary quotes and supporting documentation, only to be told that my new supervisor had arrived and that I would have to apply under him. Changing the name on the application form, I sent it in, and then decided it was probably time to go and introduce myself to this new supervisor from the United States of America.

Given the sensitive nature of the comments that follow I have changed his name to Dwayne, but all other facts remain as they happened. First, Dwayne had been appointed by the head of the SBS and was considered to be his golden-haired boy, so could do no wrong. As I was his first PhD student, Dwayne decided we should have weekly meetings – however, given that I wasn't living in Auckland (I was staying at home with Dad at this stage, rent free, as his way of assisting my project), these meetings would have meant a seven-hour round trip in my little car each week. So I suggested a compromise: a face-to-face meeting once a month (I figured I could use the trip to visit the library for reference materials, write application forms on the university computers, catch up with possible sponsors and meet with Dwayne all in one fell swoop), and a telephone meeting the other three weeks.

However, Dwayne insisted that the meetings be weekly, in his office, every Tuesday at 10 a.m. To which I replied that if the orca turned up I would be going to see them, as they were far more important than anything he might have to say in any meeting. I also took the opportunity to inform him that I was heading back down to Antarctica in a few months, and that as there weren't any regular flights he shouldn't expect me at any weekly meetings then either. His response was that I would not be permitted to go to Antarctica. 'Like Hell!' was my reply, and I left his office fuming, as I am sure he was; and so the tone was set for the rest of our encounters. Sure enough, when it came time for the very first weekly meeting, the orca *did* turn up, and I was out on the water collecting data. I still believe he thinks I faked the call-out.

The next issue that arose involved my application for funding for a boat through the university. It had been approved! I had even found exactly the boat I wanted and would have enough money from the grant to get it, the trailer, some safety equipment and a few charts. I was in seventh heaven. I phoned up the guy who owned the 4.8-metre Naiad I'd found and told him I'd take it if he could arrange transport as I still didn't have a vehicle to tow the boat home. A price was agreed upon, the boat was put on a train and sent my way, and arrangements were made for payment.

But when I took all the appropriate paperwork to the university's funding department, I was told I would have to get my supervisor to sign it. I was horrified – why did *he* have to sign it? It was *me* that had gone through all the application processes, the funding was for *my* research project on orca – but that was the system, and as much as I would have liked to buck it, I needed to follow it on this occasion. So off I set, and duly asked Dwayne to sign away. Predictably he refused. He stated that *he* wanted to get a boat which didn't have a rigid hull. *He* wanted a complete inflatable. He said that if I got a smaller engine, along with a fully inflatable boat, I could deflate the boat and put everything in the back of a van or car, and therefore wouldn't need a trailer or four-wheel drive, and could therefore cut my project costs. I was outraged. What did this guy know about boats? Practically nil, as it turned out.

Fair enough, a smaller, fully inflatable boat with a smaller engine would cost a lot less and fit in the back of a vehicle, but a boat like that wasn't going to be suitable for an orca research project conducted around the New Zealand coastline – it wouldn't be safe to travel more than a few hundred metres offshore in such a craft. The sea conditions I could expect to encounter in many of the areas I knew orca had been seen would be way in excess of what a small boat like that could handle. It would entail either extreme danger or curtailing the project to an impracticable extent. So we argued. And we argued. And eventually I informed him that I was going to seek a higher input into the situation, and stormed out.

Finally, after more than a fortnight of getting the run-around, I found the right person to talk to and explained the situation. The long and the short of it was that Dwayne was made to see the light and told to sign the form. By now the guy who owned the boat was hopping mad. I had told him the money would arrive the same day as the boat – that had been three weeks ago. I had the boat sitting at a friend's place, but still hadn't paid for it. In due course his money was delivered and I wasn't sued, but it wasn't a happy situation. And things back at the university weren't much better.

In fact, they got so bad I ended up applying to the students' association to help me have Dwayne removed as my supervisor. By now it was just over a year since I had enrolled, but I had only seen orca a couple of times. I knew the project could work – but not if the boat wasn't available all the time, being diverted to other projects which Dwayne had decided would benefit from its use. Also, I knew I would need a lot more funding to keep the project running. I was working on applications in every spare minute, but too much precious time was being taken up with trying to fight Dwayne and the university system.

Then, in the middle of all this, my mother died in a car crash. She and Dad had been divorced for a number of years and she'd been living at Waipu about one hour's drive from his house. I'd often stayed there for weeks on end as her house faced out to sea and it was possible to look for orca from her verandah. Her death, on top of everything that was happening at the university, was the final straw. I decided to pull out and chuck the whole project in.

My sister Monique and I realised that we would have to sell Mum's house to pay off the mortgage and outstanding bills. We also had to sort through all her possessions and decide what we were going to do with them. It was a very traumatic period and I just couldn't find either the time or energy to notify the university that I was dropping out. When I finally got around to it I decided to go and see John Craig, and tell him before I filed all the paperwork. We had a long talk, during which he agreed that things hadn't gone the way they should have at the university. He emphasised that I was a client of the university – that I had paid my fees and they should be *helping* me to get my PhD, not *hindering* me. John suggested that I hang in there for a little while longer and try to just let sleeping dogs lie. Who knew what might crop up? And, of course, he was right.

After my sister and I had sold Mum's house we had some money left over. Monique decided to put her share towards a home, and I decided to buy the Naiad from the university (I would have a silent partner in my mum). The funding agency agreed, the university agreed, and to my surprise even Dwayne agreed. So I walked away from all that ugliness finally ready to begin the real research part of my doctorate. I had a boat that I could use at any time, I had a camera, and now all I needed was a truck and some orca.

In a final act of closure I succeeded in having Dwayne removed from my supervising committee and John Craig reinstated. As John had just transferred to a new university department, Environmental and Marine Sciences, I became their first PhD student. Despite all the setbacks I was delighted to be back in the game and was raring to go.

Throughout all these dramas I had still been applying left, right and centre for sponsorship for a four-wheel-drive vehicle. Our neighbours, Keith and Judy Beardsly, had suggested I contact Golden Bay Cement where Keith worked. After a series of meetings and over a year of negotiating with the company executives, Golden Bay Cement agreed to sponsor a four-wheel-drive vehicle – a small truck with a cab for passengers and a covered tray for all my research equipment. The vehicle would be signwritten with company logos, pictures of orca and my orca project logo. I would have the use of it for a year, including any fuel costs associated with the research side of the project. I could also use it for educational purposes such as talks to schools, and after that year they would reassess the sponsorship.

As soon as I get a call about an orca sighting, I hook on my boat and trailer and head off. It is a bit like being a doctor on call at all times. I never know where I will be going, if I will get lucky enough to find the orca or, if I do find them, who it will be. Sometimes I have travelled three days non-stop to get to a sighting, but typically I travel about five hours to get there, and then have the same trip home at the end of a long day watching orca. **Ingrid Visser**

There was no stopping me now. I was giving an average of four talks a week – to schools, dive clubs, tramping clubs, yachting clubs, fishing clubs, the coastguard, lifeguards, boating clubs and organisations like Lions and Rotary. You name it – anyone who might own a boat or go to the beach and therefore might have the chance to see orca was worth talking to. I gave radio interviews and was interviewed for newspaper and magazine articles. I wrote articles myself and sent them off to every fishing, diving and boating magazine I could find. I wanted to get the message out there that I was looking for orca and for anyone who saw them to get in contact with me. I was an advertising machine. I was working at this project eighteen hours a day and consistently dreaming about orca as well. Every appointment I made was on the proviso that if I didn't turn up then I had heard about orca, so I was off trying to locate them. Most of my friends had completely given up on me to show for social engagements, although one, Tanya Jones, who was involved with Project Jonah for many years, also became more involved with my orca research as time went by, and eventually had an orca named after her – TJ for Tanya Jones.

By this stage my orca photo-ID catalogue had grown substantially. I had seven groups of orca, with over forty members. Things were humming along nicely and I was getting a reasonably frequent number of re-sightings of individual orca, including Nicky and Corkscrew, the two who had been my first matches. The very first orca in the catalogue, who'd been given the number A1 by virtue of belonging to the first group of orca I had catalogued, called A pod, and was the first animal in that pod, had the top of her dorsal fin completely missing. The earliest record I had of her came from some old film footage shot by Simon Cotton, an Auckland doctor who'd figured that it would be great to make a home movie about orca in the wild and had amassed a huge amount of material. He kindly let me look through all the cut and uncut tapes, and in among this I found a sequence showing A1 swimming under the Auckland Harbour Bridge sometime in 1977. She had the top of her fin missing then, and was fully grown. Using this information, and assuming that New Zealand orca had a similar life history to those studied off the BC/WS coasts, I worked backward: if she was an adult in 1977, and reached full size at around twelve to sixteen years of age, she would have been born around 1965 – so she was about the same age as me, making her at the time of writing this around thirty-eight to forty years old.

Because of her distinctive fin A1 had been photographed by the public a number of times. People would bring their photographs along to the talks I was giving and so I started to plot where these individual orca could be found and who they had been seen with. A1 was always seen with an adult male orca, Olav, catalogue number

A3. Olav was a big orca – probably the largest in the New Zealand population. I had photos of him from early as 1987 and from both the North and South islands. Olav wasn't as obvious to the general public as A1, although he did have a large notch out of the trailing edge of his dorsal fin. After a while, if I found a photograph of Olav or came across him myself, I expected to see A1. This certainly seemed logical given the evidence that was coming out of British Columbia and Washington State at the time. The orca there were to be found in stable social groups, and could be anticipated to remain that way.

So I continued to label all the different groups I came across alphabetically. There was B pod (which contained Nicky and Corkscrew) and C pod, which contained an adult male with a large notch out of the very top of his fin, who was therefore called Top Notch and given the catalogue number C1. There was a female orca with the top of her fin so mangled that it looked like it was a jigsaw piece – so naturally she got the name Jigsaw, and I gave her pod the code J, to make it easy to remember. And so it continued.

All this went well for two years. Then I started to get some conflicting data. Early on in the fieldwork I had photographed one orca called Double Notch (because of two notches out of the trailing edge of her dorsal fin) with Nicky, so she *should* have belonged to B pod. I had given her the code B10, so what was she now doing with Rocky and Spike, two adult males of D pod? Perhaps this was a meeting up of pods similar to those I had read about off BC/WS? But in those cases *all* the orca from both pods had been present. Here it was just one female switching groups. Maybe I hadn't recorded the data correctly or somehow got the photographs mixed up? Or had there been a mis-match on the photo-ID?

So I went back and double-checked everything. It all seemed right, but something still wasn't falling into place. Perhaps this female had originally been born into D pod and had just been hanging out with her boyfriend in B pod? That could make sense, and now maybe

A1, a female New Zealand orca with the top of her fin missing, presumably from an entanglement with a fishing line. She was first recorded in 1977 and is still going strong.
Ingrid Visser

she had mated and was returning to her natal D pod? But then I found that Sergeant (so named because of some stripes on his fin) had also switched allegiances. He was originally the sixth member of D pod so was catalogued as D6, but now he was spending a lot of time with B pod; but Sergeant wasn't there every time I saw B pod. *Now* what was going on? This was a sub-adult male changing groups so the behaviour was unlikely to be linked to breeding as I had thought for Double Notch. It *kind of* made sense that a reproductively active female might change, but why would a sub-adult male, who wouldn't get the chance to mate for many years, spend some (but not all) of his time with another pod? I was completely confused. And then things started to get really crazy.

Every time I saw the orca, and no matter who it was, it seemed they had switched allegiances somehow. It might have just been for that one encounter, but it sure was making a mess of what had previously been a very tidy system of cataloguing. Eventually, after spending nearly a year trying to sort out what was going on, I decided there was only one thing for it, and that was to scrap the whole system and start the catalogue right from scratch again. This was going to be a huge job, but some orca now had codes that no one but me would ever be able to understand, such as DB10a – this orca was first seen in D pod, and was the tenth member of that, but later switched to B pod; but there was already a B10 so this 'new' B10 had to have an 'a' suffix. As I had always anticipated that this project would be ongoing well after I'd died, I wanted to be sure that the catalogue was manageable for whoever took it over. If it was going to be changed it would have to be now.

So, grabbing the bull by the horns, I dumped all the old catalogue numbers and started afresh. From now on all New Zealand orca would have the prefix NZ and this would be followed by a number. This meant that no matter who the orca typically hung out with, they would all be numbered independently. It would also mean that pod affiliations would no longer be apparent from the catalogue number, as was the case off BC/WS. But given that these orca were chopping and changing between pods, and looked like they were going to continue to do so, it seemed the logical approach. So, although the orca retained their nicknames, their catalogue numbers all changed. Therefore, A1 and Olav became NZ1 and NZ3 (A1's calf, Zoely, was NZ2), Corkscrew who had been B1 and Nicky who had been B2 were now catalogued as NZ15 and NZ16, and so on. They still had their names and they were still the same orca, but within my New Zealand Orca Research Identification Catalogue they were filed consecutively. Now the fieldwork could continue and the data entry would be considerably smoother.

One day in early 1994 I was out with Rocky and Spike and other orca, including Double Notch and Venus (who I thought might be Rocky's mother) in the Bay of Islands, and I followed them as they headed into a shallow estuary. Although all the popular and scientific literature to date had told us that orca were open-water cetaceans and shouldn't really be found in these shallow areas, I had been working with New Zealand orca for about a year with my own boat and had witnessed this behaviour a couple of times before. From my previous observations I could expect the orca to be in the estuary for most of the high tide and that they would be hunting.

Sure enough, the orca continued to head into the estuary and before long they were in water that was so shallow their dorsal fins were sticking out. They split up from their tight travelling group and spread out over the estuary. Suddenly there were orca speeding about all over the place and in very shallow water. They were obviously chasing something, but the water visibility was zero because of all the stirred-up sediment, so I couldn't see what they were hunting.

Venus was hovering in water about 3.5 metres deep with her tail waving in the air and appeared to be trying to dig in the bottom. She lost her balance and tipped over backwards, then surfaced right next to my boat with her rostrum (snout) covered in mud all the way back to her eyes. She was digging all right, and whatever it was she was after seemed to be trying to escape into the mud. After watching her digging for several more minutes I saw a huge billowing of bubbles and then she surfaced again. This time she had something in her mouth and it was still alive – a stingray. The ray was large enough that its wings stuck out each side of her mouth and they were flapping. Its tail was also sticking out of her mouth and was whipping around trying to sting her with the barb. Venus suddenly tossed her head and flicked the ray like a frisbee. It flew through the air and landed upside down on the water, remaining there motionless as Venus quickly swam over to it and grabbed it again. By now Rocky was well aware of what was going on and had headed over to join her. Between the two of them they pulled the ray apart and spent about ten minutes sharing their lunch.

At the end of the tide the orca headed out of the estuary and back to sea. I had seen them take fifteen rays that day. This turned out to be more than anyone else had ever recorded, anywhere in the world, *and* at any time in the whole history of orca research. But I didn't know that as I continued to observe orca chasing rays around the New Zealand coastline.

On another occasion later in the year I watched them in Whangarei Harbour where the water was much clearer than in the Bay of Islands estuary. Here I could clearly see the rays racing desperately across the shallow sandbars with hungry orca hot on their tails. The rays did their best to avoid the orca, sometimes heading under the wharves where the orca couldn't pass between the piles, and sometimes going right up onto the beach to avoid them. I even saw one ray practically climb a rock wall in its efforts to get out of the water as an orca calf pulled on its tail. The orca were chasing rays into water that was so shallow they got stuck, having to thrash and roll around before they could finally get off the sandbars to resume their pursuit.

In November 1995, Dr Robin Baird, an orca researcher from the BC/WS area, was out visiting some other cetacean researchers in New Zealand, and I sought his advice on how I was conducting my research. Although I had been following the orca for a while now, I had yet to actually work with any other orca researchers and was concerned that I might not be gathering the data in the right way, or perhaps missing important aspects of their behaviour. After spending just a few hours on the water with me, Robin's advice was clear – the orca I was watching were unique and I should scientifically publish my data as soon as possible.

So I set about writing my first paper. Originally it was fairly rough and required quite a lot of editing but eventually I got it published in *Marine Mammal Science*. Called 'Benthic foraging on stingrays by killer whales (*Orcinus orca*) in New Zealand waters', it described the orca behaviour as digging in the bottom of the sea for rays. As it was published in a US journal they insisted I use the term 'killer whale' as that was their common name for the species. There was a lot of additional data and information that I wanted to include in the paper, but much of it was edited out by the reviewers during a lengthy and involved process. I had submitted the first draft to the journal in September 1997 but it didn't get into print until January 1999. Regardless, I had published my first scientific paper, and I was finally getting a chance on the world stage to show that the New Zealand orca were unique.

Back home, ordinary New Zealanders were becoming much more aware of how special our orca were too, thanks to my blitz in the media and all the talks that I was giving. The public were phenomenal in their responses and attempts to help me locate orca. I would get calls from people out picnicking at the beach, or someone out fishing with their kids. People crossing the Auckland Harbour Bridge to work in the morning would call me from the middle of the traffic. Commercial fishermen, the water police and ferry captains all contacted me. People who had houses overlooking the water or who had gone on a dive trip and seen orca all got in touch. Kids on school trips who saw orca , and pilots flying over harbours, called as well. It was incredible feedback and I appreciated the contribution made by all of them to the ongoing study of New Zealand orca (and still do).

During this time the whale- and dolphin-watching industry had been growing rapidly, and I was receiving more and more calls from different companies which had become established around the New Zealand coastline. One outfit up north wanted to set up a new boat dedicated to taking people out to watch cetaceans, and this was how I met Jo Berghan, who wanted to work specifically with cetaceans and was helping the company set up their education programme. She was going to concentrate on the bottlenose dolphins found in the area (*Tursiops truncatus*) but would record data on any other species that she came across. Jo began to collect some great photo-ID shots of orca as well as sighting information, and she and I were out working together one day when we came across two orca – an adult female and male – and witnessed an amazing event.

This turned out to be the smallest group of orca I ever encountered during my studies of them in New Zealand, and it may have been just the two of them because the female was about to give birth. Not twenty minutes after we had arrived we suddenly saw a third member of the group enter the world – an orca so tiny it resembled a wind-up bathtub toy. It was popping out of the water like a cork and just didn't have any coordination when it came to surfacing. It was scurrying around on top of the water like a little insect, so I named it Kootie after the colloquial name for little creatures that scurry around in your hair, also called nits. Kootie was just so cute – with its tiny yellow eye patches which were typical of calves six months or less in age. The yellow colouring is thought to signal to other orca who don't know the calf that it is just a baby, and to give it some leeway with any errors that it might make, or to take special care. (In the BC/WS area the researchers have found that orca calves tend to be born between October and March, the Northern Hemisphere autumn and winter, but so far I haven't identified any seasonal trends for the breeding season of the New Zealand orca.)

One of the theories about the black-and-white colouration of orca is that it helps the animals to coordinate when swimming. Certainly, from underwater the white eye patches and flanks are highly visible when they are swimming alongside each other. Perhaps the white flanks also help to guide the calf to where the teats are. Regardless of the ecological reason for their colouring, orca look stunning, and in the case of wee Kootie, very endearing as well. After only a few hours Kootie had managed to get the hang of the swimming thing and was happily accompanying her mother as they headed out of the Bay of Islands.

Kootie's mother, Hollowhead, had a large hollow just behind

her head which looked like a spinal deformity. When Kootie was born I thought perhaps the shape of the deformity was just a expression of the pregnancy, but no, even years later Hollowhead still has that large dent. There have been other orca in the New Zealand population with spinal deformities and one of those was called Marilyn (after Marilyn Munroe), as she had a deformity which gave her caudal peduncle (the area between the dorsal fin and the tail flukes) a big, permanent curve. As Jo had seen some similar types of deformities in bottlenose dolphins we decided to publish a scientific paper about these anomalies to let others know about our observations. It turned out to be an interesting exercise as we discovered ten individuals from three different species of cetaceans with similar deformations of their spines. We were unsure what caused the spinal problems and most likely never will know, but for some other types of injuries we were acutely aware of their origins.

An orca which had suffered some horrendous injuries was first photographed by me in January 1995, in a series of beautiful flooded valleys of the Marlborough Sounds at the northern end of the South Island. (These valleys had been flooded for centuries and now provided prime habitat for rays, and therefore orca.) While out following a group one day I noticed one orca who from a distance looked different. It appeared to be a very large calf, as the dorsal fin was extremely falcate (curved, like a dolphin fin), which is characteristic of orca calves. However, that shape is typically lost as the orca gets older; and in the case of some adult female orca the dorsal fin changes shape so much that it may be triangular and shark-like in shape; adult male orca may have a dorsal fin that can be two metres high. But this animal's dorsal fin was half the size of a typical adult female orca fin, though the orca itself didn't appear to be a calf. As I got closer I could see that she was just slightly smaller than my boat and had three large V-shaped scars on her caudal peduncle.

Prop, who is now catalogued as NZ25, had obviously been hit by a boat's propeller – hence her name. Generally, only two of the

This female orca has been hit by a boat propeller, so she got the name Prop. She has two easily visible cuts in her back, and a third further down towards her tail. She has a very small dorsal fin, and is a third smaller than other orca her age. I suspect that she was hit when she was only young and put all her energy into recovering, rather than growing. To have survived, she must have received a lot of help from the other orca in her group. **Ingrid Visser**

cuts were visible at any one time, as the third was further down her tail and usually underwater. Each cut was deep enough to be clearly seen and although the wounds had healed large chunks of flesh were missing. When viewed from the side these cuts appeared to penetrate almost to the spinal column; the one closest to her dorsal fin extended right into the saddle patch. When viewed from the front the depth of these cuts was just as graphic, giving Prop the appearance of an armadillo.

Using the same theory to age Prop as I had for A1, I worked backwards from the earliest photograph I had found of her, 1982, and assumed that she was born around 1970 to 1976 at the latest. Given this age, and based on the findings from the PNW orca, Prop should have been around five to six metres in length, but she was only about 4.5 metres long. Because of this small size, and in conjunction with her small falcate dorsal fin, I came to the conclusion that Prop had most likely been hit by the propeller when she was a small calf and subsequently had put all her energy into healing the extensive wounds and very little into growing larger. (Studies of other injured wildlife have found that injuries increase energy consumption significantly; add the fact that the animal is often not able to hunt or may not feel like eating, further reducing energy levels, and even less energy is available for growth.)

One study suggested that cetaceans that deal with boats frequently do so with a possible loss of energy which would otherwise be used for general living and reproduction. Therefore, if they have to constantly expend energy avoiding vessels, they may ultimately lose out on the game of life. This is a problem for cetaceans in all areas of the world where there is heavy boat traffic, and particularly for the orca off the San Juan Islands and Vancouver Island, which straddle the Washington State and Canada border. A real-life example of this phenomenon occurring here came about from studying Prop's group long term. After watching her for many years I had amassed quite a collection of photographs. I could plot where she had gone around the New Zealand coastline and when she turned up in different locations. I found that she and her group were travelling about a third less distance per day compared with other orca in the New Zealand population. It was possible that her group was travelling at a slower rate to accommodate her injuries, which although now healed have lasting effects on her health.

Cetaceans may eventually become used to boats, but this may not prevent them from becoming damaged by propellers. In the end, it is the responsibility of the boat driver to take into account that animals, especially young, inexperienced ones, may not associate boats with direct physical injury. This aspect of orca life has now become a standard part of the talks which I give to the public. I am in the process of producing a leaflet outlining safe boating practices around orca (and other cetaceans), using some of the graphic photographs I have collected of New Zealand orca which have been hit by small craft. If these animals had been hit by large ships they surely would have died, but regardless of the size of the boat, I want to try to prevent any further injuries. However, boats are not the only threats encountered by New Zealand orca; some of their preferred food sources can also inflict considerable damage if not tackled carefully – but first, let me put New Zealand orca into a worldwide framework.

# CHAPTER FIVE
## Orca around the World

The waters of the Pacific Northwest (PNW) – off Alaska, British Columbia and Washington State – remain the hub of world orca research. It is the area of both the longest-running and also the highest number of projects (around twenty different projects concurrently). This has meant that most of the early information known about orca and most of what we continue to learn about them, in terms of finer details, comes from scientists working on those projects. Their results continue to influence the thinking of orca biologists now working in other areas.

From the long-term photo-ID projects conducted in the PNW we first came to realise that orca were very long-lived – female orca have a maximum longevity of about eighty to ninety years and an average life expectancy of around fifty. Male orca, on the other hand, have a maximum longevity of about fifty to sixty years and an average life expectancy of around twenty-nine. The studies from which this information came were based on two populations of orca collectively called the 'Residents', as these orca were known to spend a considerable part of their lives in the area. The two communities were dubbed separately the 'Northern' and the 'Southern'.

It turned out, contrary to initial assumptions, that the groups the scientists were observing were based around a matriarchal society – in which an adult female was the group leader and all the descendants were traceable along a female line. Originally these matrilineal theories were based solely on observations but were eventually backed up genetically from biopsy sampling. The 'Matrilineal Group', as it was termed, may contain up to four generations of orca, and its individual members are usually seen right next to each other – in many cases even surfacing in synchronicity with each other, particularly when resting.

When two or more Matrilineal Groups consistently travel together they are termed a 'Subpod'. Subpods may be comprised up to eleven Matrilineal Groups, however most contain two. It is thought that the females within a Subpod are likely to be closely related – such as cousins. When a number of Subpods prefer to travel together – although they may spend weeks or even months apart at times – it is termed a 'Pod'. When a number of Pods get together it is called a 'Superpod', although this may happen for only a few hours or days at a time. Two further classifications for PNW orca are the 'Clan' – which is a collection of Pods which have similar dialects (and are most likely related to each other, based on these vocalisations) – and the 'Community' – any Pods which have been observed together at least once during the study.

Change within the delineations of orca groups – albeit slow, given the longevity of the females – might come about because of the death of a matriarch of the Subpod – perhaps triggering the rearrangement of Matrilineal Groups into new Subpods. If no reproductive females survive within a Matrilineal Group the line will die out when the last member dies, and the Subpod will be comprised of one less line.

One very interesting feature which scientists noticed while looking at all these relationships was that animals never moved between groups. The only way an orca became a permanent member of a group was to be born into it, and the only way it left was when it died. This lack of dispersal (for both sexes) from the natal group has not been documented for any other species of mammal, making orca unique in the world in this matter. Of course, if orca never leave their group, this raises the issue of inbreeding. In the case of the PNW orca, however, we know that outbreeding does occur and this has been backed up from genetic analysis of biopsy samples.

The researchers have also found that although an adult male orca may be sexually mature at fifteen years old he isn't physically mature until around twenty-one, and it may be many years after this before he gets to father a calf. We still don't know, however, who chooses whom. But it is probably the female who chooses the mate, given the outsized secondary sexual characteristics – large dorsal and pectoral fins, and curled tail flukes – of adult male orca. Sexual selection is driven by the female in other species of the animal world (e.g. peacocks). However, adult male orca do have some physical constraints on their secondary sexual characteristics; if their dorsal or pectoral fins got too large they wouldn't be able to turn effectively (and in the case of New Zealand orca, manoeuvre in very shallow water), and if their tail flukes became too curled they wouldn't be able to swim efficiently, therefore making it difficult to hunt and to avoid predators.

Female orca also have a longer breeding life span than males, typically giving birth to their first viable calf at around fifteen years of age, and going on to produce an average of just five calves during a twenty-five-year reproductive life span. Once the female is no longer producing calves she doesn't die, but stays within the group. Studies of pilot whales (which are closely related to orca) have shown that females who do not have calves can still produce milk to feed calves which are not their own. Perhaps post-reproductive female orca, some of which may be around for a further twenty years after they have stopped breeding, are helping young orca to catch food, or training them in other survival skills?

One possible explanation for the longer life span of female orca is the effect of reproduction on some of their bio-accumulated poisons and toxins. As these contaminants are fat-soluble, when a female gets pregnant she passes on some of her own personal collection of these problem chemicals to the foetus; this is termed 'offloading'. Then when the calf is born it is fed milk which contains over twenty per cent fat, and therefore twenty per cent of the milk will contain a further offloading of the mother's bio-accumulated chemicals, which may contribute to the very high level of mortality among calves – about forty-five per cent. While offloading might contribute to the longer life span of the female, it is also possible that the shorter life span of male orca is linked to these pollutants – as they don't get a chance to offload some of these accumulated chemicals, which are known to affect their immune and reproductive systems, and most likely the rates of cancer and other malignant diseases, in other mammals.

Based on the findings from the PNW long-term studies and combined with the small numbers of orca known to exist for each population, the Canadian government recently declared their two Resident communities of orca as 'Threatened' and 'Vulnerable'. Ranking animals according to their populations, and the threats associated with those populations, also has a standardised version across the world, known as the 'Red Data List', produced by the International Union for the Conservation of Nature (IUCN). The first one for cetaceans, published in 1991, used seven different categories to describe their possible status and listed orca as 'Insufficiently Known'. This ranking inferred that orca were *suspected* (but not definitely known, because of lack of information) to belong to one of the Endangered, Vulnerable or Rare categories. Since 1991 there have been two additional IUCN Red Data Lists (published in 1996 and 2000); however, the 1991 report remains the only Red Data List specifically for cetaceans.

The categories for all animals have been modified since the 1991 listing, and as of the 2000 version, orca worldwide are now designated as 'Lower Risk, Conservation Dependent'. This is complicated, but essentially means that where orca are the focus of a continuing orca-specific or habitat-specific conservation programme, should that stop it would result in the orca qualifying for one of the threatened categories within a period of five years. However, conservation programmes, or acknowledgement that these programmes are required, are in place for only a limited number of orca populations, such as the Canadian ones.

Unfortunately, in order for many conservation measures to be taken, a species or population often must be ranked in one of the threatened categories before funding or conservation resources will be allocated. And while we know some details about the orca found off the PNW coastline, we don't even have a global population estimate. The IUCN report is very inconclusive and refers to sightings in different areas in terms such as 'a few thousand', 'sporadic', 'frequent', 'clusters', 'occasional sightings', 'concentrations', 'regularly seen', but it does list some numbers, where known. Adding up all the actual numbers listed, the world population is estimated at 70,000 orca south of 60°S (basically, the waters around Antarctica) and only 7367 individuals north of this latitude (i.e. the rest of the world). Personally, I think the number given for Antarctic orca is unreasonably high, in light of the findings in other areas of the world.

There are still plenty of places around the world where orca are known to be found, yet are not studied – such as Australia, Namibia, South Africa, the east coast of North America, Japan, Chile, Peru, the Indian Ocean and most parts of Antarctica, to name just a few.

So why aren't they being studied there? Many orca live far from shore in rough open ocean where it is difficult to conduct research, but the thing that primarily holds back orca research is funding – in other words, money is the key to these ventures.

However, where orca *are* being studied in detail we are finding that there are a lot fewer animals than originally anticipated. For instance, in the PNW where some people initially thought there were thousands of orca, there turned out to be fewer than 300 'Resident' orca and about 400 'Transient' animals living along the whole coastline (with a further group of orca termed the 'Offshores' with an unknown population size). The Transient orca were first studied in detail by Alexandra Morton during the mid to late 1980s, and then by Robin Baird who took up the cause in the 1990s. The Transient orca were found to differ markedly from the Residents (which fed chiefly on salmon), specialising in feeding on marine mammals – primarily seals and porpoises but also dolphins and larger whales – and tending to move through an area fairly swiftly, as their name implies. Many other things were observed that didn't fit into the model which had been proposed for the Resident orca, including the way they hunted – typically silently and in small groups (between one and six orca) – while hugging the coastline in search of their prey. It was also found that Transient offspring may or may not stay with their mothers. This was a major breakaway from the thinking of the time and helped other orca scientists realise that the established model for PNW orca didn't have to be the only answer.

By the early 1990s studies of orca had begun in other areas of the world. Primarily these were off Norway, the Crozet Islands (in the southern Indian Ocean), and my own project off New Zealand. Additional projects, which would turn out to be short-term or intermittent, were also initiated in Antarctica, Argentina, Brazil and Iceland. All of these started out with photo-ID techniques and branched into different aspects of enquiry, depending on the researcher, the population of orca, the facilities (and funding) available, and the questions being asked.

Off Norway a few researchers were working independently with orca and they all based their work on earlier observations made for the whaling industry. Working with slightly different groups of orca, in slightly different areas of the coast, they all found the same thing – the orca were feeding on herring in a method that two of the researchers, Tiu Similä and Fernando Ugarte, came to describe as 'carousel feeding'. The herring were rounded up by the orca who worked cooperatively to bunch the fish into a tight group. Individual orca then took turns at swimming in close to the herring and striking the fish with their tails. The orca would then pick up the fish one by one, which were stunned either by a direct hit or by the force of the movement close to them.

In another study off Brazil orca were found to be taking fish off longlines set by fishers. Longlines either hang vertically in the water and are usually around 200 metres long, or consist of a groundline that lays along, or near, the sea floor; it can run for ten kilometres or much longer. Each groundline has sixty to 100 hooks spaced about a metre apart on side lines; each hook is expected to catch a fish, so for a ten-kilometre stretch of line a fisher might catch up to 10,000 fish. Now this is unlikely, but assuming the fisher might get even half this number of fish that is a lot of fish concentrated in one area. And the orca learned this very quickly. The fishers off Brazil were pursuing swordfish and tuna – both fast-swimming prey which would normally be at the outside edge of an orca's swimming speed (we know orca can swim at speeds of over twenty knots, which is about thirty-five kilometres per hour), but they might catch them often enough to get a taste for this food. However, if these fish are already caught on a line, why not just wait until the fishers bring them to the surface and take them then? And this is exactly what the orca did, turning up around the longlines only when it came time to haul the lines in, and never bothering to hang around while the

A juvenile orca removing a small school shark off a hook by taking the shark by the tip of its tail and pulling until it 'pops' off – thereby preventing entangling itself in the hooks. The fisherman who took this photo asked to remain anonymous.

lines were being set, or while there were no fish on them. The orca got so good at taking the fish off the lines and not getting caught on the hooks themselves that they were taking virtually all of the catch, prompting the fishermen to start taking shots at them – not a good scenario for either group.

Similar situations have arisen all around the world, such as off Alaska, where orca have been documented staking out longlines time and time again. Researcher Craig Matkin found that the rate at which the orca take the fish from the lines only increases and devices such as tuna bombs (underwater dynamite) do little to stop them. In the waters off New Zealand I have seen orca taking fish off longlines and so far most of the fishers are pretty okay with it, but as the orca here are only taking five to ten per cent of the catch, I suspect that when they take more the fishers will become less tolerant and we will see more shooting at orca as a result.

Overall there does tend to be a strong research focus on what orca are eating and the dramatic and often prolonged way in which they hunt and attack their prey (collectively termed 'foraging'). Our first real glimpses into what orca were eating came from the examination of the stomach contents of animals which either stranded or were caught. We know now, based on these early records and recent observations of foraging behaviour, that orca take an exceptionally wide range of prey which includes jellyfish, squid, sharks, rays, bony fish, sea turtles, birds, seals and other cetaceans. There have also been records of orca taking a few land mammals, such as moose, and recently reports have started to surface of orca taking bears (but all of these attacks so far have been recorded when the land mammals were swimming). From our collective studies around the world it does appear that although orca take a wide range of prey, individual orca tend to specialise in a smaller category of food – for instance, eating just fish or just mammals, but rarely both.

Other orca behaviours, apart from foraging, also feature in studies, although because of research constraints it tends to be those which can be observed from the surface which receive the most attention. These might include social behaviours such as food-sharing (which ties in nicely with the foraging behaviour); some studies have concentrated on who shares food with whom (are the orca related, or just good friends?), and teaching behaviours – again, often associated with hunting techniques. The fact that so many of these behavioural aspects are interlinked and can be studied either separately or as part of a whole is a challenging but fascinating side of working with orca. Group size is a prime example of this interlinked behaviour.

We have found from studying how often an orca hunts (and actually catches and eats the prey), what type of prey it was (a small fish compared with a big fish, or a dolphin compared with a whale), and whether the orca shares the food, that group size is intrinsically linked to foraging. When prey is very abundant but small (for instance, herring), orca group sizes also tend to be large, such as those found off Norway where the orca cooperatively herd the fish into tight schools to prevent them escaping. If orca are hunting for prey which is more 'street-smart' than your average fish and isn't found in large groups, then perhaps it makes more ecological sense for the orca to operate in smaller groups in which each orca has a set task in the hunt. A prime example of this *modus operandi* is a group of orca who hunt for dolphins off the small whale-watching town of Kaikoura in New Zealand. There are four orca in this group, two adult males and two adult females. They have never been seen with any other orca and always travel together. These orca have only ever been seen feeding on dolphins, and have the technique pretty well sorted out. I have never seen them miss once the players are all in place and the hunt is on.

Their method works along the following lines: the orca travel in 'front-formation' – four abreast, spread about touching distance apart – and cruise nonchalantly towards a small group of dolphins (they never try this method with more than about ten dolphins). The dolphins are very aware that they are on the menu and head away, but not too fast, as they don't want to draw the attention of the orca just in case they aren't really hunting. After following the dolphins for anywhere up to thirty minutes one of the female orca doesn't surface with the others (and isn't seen surfacing anywhere else). This orca, which I have come to call 'Stealth', still doesn't surface the next time the others rise to the surface, or the next. In fact, typically Stealth isn't seen for about ten minutes before the three remaining orca make their move – and move they do.

They take off towards the dolphins at high speed, which is incredibly dramatic as three adult orca begin hurtling along the surface like hunting machines, and the chase is on. The dolphins are fleeing for their lives, and they know it – they fly out of the water and don't even seem to touch down before they are off again. The three orca are still behind them, and closing fast, but suddenly from the front of the dolphin group one of the dolphins goes flying into the air having just been hit from below by Stealth as if it was a tennis ball – tumbling through the air as it turns somersaults. Just before it reaches its aerial apex Stealth is right below it – also hurtling through the air from her follow-through. She grabs the dolphin in mid-air, then falls back into the water with it in her jaws and proceeds, with the other three orca, to devour the meal.

This vivid example highlights the need for a small group of orca to coordinate such an attack. Each orca must have learned its role in the ambush process, and the fewer orca involved the less chance things can go wrong. But, of course, there needs to be enough orca to ensure that the dolphins will head off in the right direction, and enough to turn them should they try to escape in the wrong one (and possibly enough orca so that the dolphins don't necessarily notice when one of them goes 'missing', but not so many that the dolphins could detect them from a long distance away and be able to avoid them altogether). Once the dolphin has been killed the orca would want, after having put so much effort into a hunt, to be sure that they got enough food – so small group sizes make ecological sense in such a scenario.

On a limited number of beaches on the Patagonian coast of Argentina, a population of orca have been recorded taking sealion pups from the beaches. The orca use deep water channels with very steep slopes which come right into the beaches to approach their prey. When the young sealion pups are playing at the water's edge the orca come right up onto the beach and take them. Both male and female orca hunt this way, and some are much more efficient at it than others. On one occasion (while working with Juan Copello, a researcher based in Argentina studying orca) I observed this behaviour and saw a female, called Magga, take ten pups in two hours. Magga's efficiency was impressive, as she caught a pup, took it out to the juveniles to eat and then returned to the beach to catch another one, all within the space of a few minutes. Later, when everyone had presumably eaten their fill, Magga, along with another female (who had been less successful), came into the beach with the juveniles, presumably to teach them to hunt. This teaching of such a high-risk hunting method is one of the keys to its success and it may well take years before an orca achieves any degree of capability. So again, foraging and socialising are unequivocally linked.

But just how much food does an orca require? We don't really know exactly, but from captivity studies we know that an orca eats around four per cent of its body weight in food per day – so for an average-sized female orca this would equate to about thirty to thirty-five McDonald's Big Mac hamburgers and about forty-five to fifty Big Macs for an average-sized male. However, we must take into account that living in captivity is far from normal life for an orca (and of course they don't eat Big Macs). In the wild they would have completely different energy requirements, not to mention the social aspects of foraging in which you might lose some of your food intake through behaviours such as food-sharing or providing milk for a calf. But it is also likely that orca are not eating consistently, in that they may go hours or days without food and then come across a plentiful source and take full advantage of it while they can. I have recorded such fluctuations in New Zealand waters when I have followed a group of orca for up to twelve hours and not seen them feed at all (they slept for nearly seven hours of that twelve), and then followed another group for just a few hours and seen them take in excess of twenty rays, of which more than half were shared with at least one other orca.

So if we know roughly how much food an orca requires and we have some idea of what they are eating and how they are consuming it, how much do we know about what is going on deeper down in the ocean or where these predators go all year? Even though we have the use of tracking tools such as TDRs and we know that orca

Orca spend less than ten per cent of their lives at the surface, so to get a good understanding of what they are doing, it helps to get into the water with them. Social interaction can be observed, as with these two young females who spend a lot of time together. **Ingrid Visser**

My research boat is a small rigid-hull inflatable. It is like a four-wheel drive for the ocean. I have to go out in all sorts of weather, although I won't necessarily go out to look for orca when it's rough, but coming home I never know what the sea conditions will be like. **Andrew Penniket**

New Zealand orca often come into city harbours. This photograph, taken in Auckland Harbour, shows the houses in the distance. Often people out walking their dogs or sitting having a cup of coffee on their front terrace will call and alert me when the orca are visiting. **Ingrid Visser**

This orca, known as Miracle, is very interactive. She will often come right up to my boat and swim behind it for minutes at a time. She stranded and was rescued in 1993, and was re-sighted in 1996. Since then she has been seen every year. I suspect, but can't prove, that she is interested in people because of her encounter with her rescuers. Recently she had a calf called Magic. This is the first time a rescued orca has been known to breed after stranding. **Rinie van Meurs**

Nearly a year after his stranding, Ben was hit by a boat. You can see the three propeller strikes on his body and the dramatic fourth strike on his fin, slicing right through it. The tissue is rotten and the front part of his fin is bleeding.
**Ingrid Visser**

Type B Antarctic orca are grey with a dorsal cape and very big eye patches. This one was photographed off New Zealand in May 1997 – the first evidence of Antarctic orca leaving polar waters. This sighting was disputed by some, who suggested I had mixed up my films; however, this photograph shows hills (not ice) in the background. **Ingrid Visser**

This curious young orca approached repeatedly, hanging motionless below me, slowly opening and closing its mouth. Although some mouth-opening behaviour can be classified as a threat – typically those done quickly, with a loud snapping noise – this one was definitely not an aggressive move, more one of curiosity or play. **Ingrid Visser**

A young orca chewing on my fins as I was trying to photograph other members of its group. Other divers have reported orca chewing on their fins – it appears to be a game, not an attempt to get lunch. **Ingrid Visser**

Many people don't believe me when I say I swim with orca – they think I must do my research from a shark-cage or some other protective device. It took ten years to get a photo of me underwater with the orca. Here two orca are checking me out, and the pectoral fin and side of the head of a third is in the top right corner. **Brad Tate**

New Zealand orca spend much of their time in close to the coast, and in shallow water. This photo shows just how close, with the hills in the background and the bottom of the ocean visible. **Ingrid Visser**

A sub-adult male accompanying a female, presumably his mother. If New Zealand orca have a similar social structure to some of the orca found off North America, then the males never leave their mothers. Adult males seen in these groups are the sons, brothers and uncles of the female orca. They mate between groups, and then go back home to their mums.
Ingrid Visser

Three orca hunting for rays in shallow water. The middle orca has its white belly towards the camera. The orca on the left is a juvenile and is watching the adult females as they hunt. **Ingrid Visser**

Rocky with a ray, which is draped upside down over his rostrum (snout) and still alive. Rocky gets his name from hunting rays in among the rocks and is a very efficient hunter. **Ingrid Visser**

This adult female orca was hunting in water so shallow she had to tip onto her side to get over the rocks. She was hunting for rays among the weed and rocks. New Zealand has one of the highest rates of orca strandings in the world, which seem to be related to this method of hunting in shallow water. Ingrid Visser

This photo looks down on two orca as they food-share. They have an upside-down eagle ray between them. The orca on the right has the wing tip in its mouth. Connective tissue from the wing of the ray is stretched between the two orca and its liver is the pinkish organ in the middle. Ingrid Visser

A sub-adult male swimming with me off the coast near where I live. The long fins I wear help me to swim faster, but I will never be able to keep up with the orca, who can swim at speeds of over thirty kilometres an hour. By allowing me to swim with them, I have observed behaviours never seen before. **Brad Tate**

can hold their breath for around twenty minutes, what goes on when they leave the surface is still little understood. We have only a very limited idea of where most orca populations go when they are not in the core study area, although we are slowly coming to realise that orca can travel quite substantial distances out of these locations (as evidenced by photo-ID data).

With the increasing use of the Internet, scientists are updating their catalogues more frequently, receiving more images from the general public, and exchanging important catalogue information between far-flung researchers. To date there have been no verifiable long-distance matches, such as between the Arctic and Antarctica, but that will possibly occur in the future and most likely with animals about which we have limited information so far. For instance, there is still very little known about the third population of orca found off the PNW called the Offshores, but they appear to be wide-ranging like the Transients, and perhaps feed on fish like the Residents. Physically they don't look like Transients and genetically they are more closely related to Residents; however, they have never been seen to interact with either group (although Transients and Residents typically tend to move apart and go their own ways, when it comes to that).

Another population about which we know tantalisingly little is the orca found off Antarctica. We have known for a long time that orca are found there, but the environment is extremely difficult to work in and the animals are highly mobile, as is their habitat. The continually moving ice makes the option of a land-base from which to conduct research nearly impossible. However, progress is being made and I have started to compile the first photo-ID catalogue for Antarctic orca and even have some matches. Most Antarctic orca look completely different from orca elsewhere and it seems that there might be three distinct types. One of these populations looks similar to the typical orca with which we are familiar, in that it is predominantly black and white with a small area of grey behind the

Orca around Antarctica often spy hop to investigate potential prey on the ice – but in this case the orca was looking at me on the bow of the ship. Although it looks black and white, this orca is very grey on its back, a typical colour scheme for Type C orca which are found near the ice. **Ingrid Visser**

dorsal fin, and is termed Type A. However, the other two types of orca are varying shades of grey and white and rarely have any black on them at all. The two grey types are distinguishable by their eye patches – some have very large eye patches (Type B) and some have very small ones (Type C). Before 1997 these grey orca had never been described north of Antarctic waters. Then in May of that same year I saw a group of eight grey orca with exceptionally large eye patches (Type B) off the Bay of Islands. Then in 2001 (off the Hen and Chicken Islands) and again in 2003 (off the Tutukaka coast) I saw more grey orca, and both these groups were orca with small eye patches (Type C).

Last time I saw the grey orca off New Zealand I managed to get a hydrophone recording of their calls, and promptly called world orca acoustics expert John Ford and asked him to listen to my recording down the phone line. His responses were unequivocal. He could, from a recording of the grey orca vocalisations, identify them as Antarctic orca, without my having given him any idea that they had been taped off New Zealand. I then told him where in fact I had recorded them. He was thrilled, as he had known that I had seen Antarctic grey orca in New Zealand waters before and was delighted that they had been back and we now had a recording of them. John and I hope to work together on comparing these calls with recordings of orca made in Antarctic waters to see if we can get any matches in the call types, although it remains to be seen just where these grey ghost-like orca are really coming from and where they go.

Taking into account the work of the Russians (who proposed, in 1981 and 1983, the existence of two new orca species, *Orcinus nanus* and *Orcinus glacialis*), and the grey orca I had seen in Antarctic waters and the grey orca I had seen off New Zealand, I put forward a theory at an orca conference in 2002 that pigmentation could be used to distinguish between different populations of orca. In essence I was suggesting that although worldwide the typical colour pattern for orca was black and white with a grey saddle patch, variations in this pattern had been observed, and they might be linked to distinct populations or even separate species of orca in different geographical regions. I also suggested that further non-lethal research such as photo-documentation of pigmentation patterns and genetic sampling might support the theory. It turns out I may have been right. A researcher working with the US government received enough funding to head down to Antarctica and collect some biopsy samples for genetic analysis. His results at the time of writing this are still preliminary but lean strongly towards these orca not having interbred with other orca for thousands of years, meaning they *are* likely to represent separate species.

Although there is a lot more known about orca than has been touched on in this chapter, and despite the numerous studies (which make orca one of the most well-researched and best-known of all cetaceans), there are still remarkable gaps in our understanding of these animals. For instance, we have very little knowledge about what orca do at night. Obviously they possess echo-location, so are more than capable of navigating and locating prey in the dark, but do they actually hunt, or do they sleep more? Currently we can't say. This type of contraindication of findings – knowing so much, yet still knowing so little about them – continues to provide impetus to my study of orca in their natural element.

## CHAPTER SIX
### Dexterous Predators and Dangerous Prey

By now I had discovered four species of ray which were recorded for the first time – anywhere in the world – as orca prey. These were the short-tailed stingray, *Dasyatis brevicaudatus*; long-tailed stingray, *Dasyatis thetidis*; eagle ray, *Myliobatis tenuicaudatus*; and the electric ray, *Torpedo fairchildi*. The first three species have long sharp barbs in their tails and these are used by the rays for defence. In the past I had sometimes seen rays without tails while I was out diving and had always assumed fishermen had cut them off, but after I began my research I started to realise that perhaps orca were also to blame by biting them off.

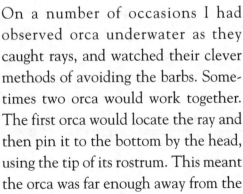

On a number of occasions I had observed orca underwater as they caught rays, and watched their clever methods of avoiding the barbs. Sometimes two orca would work together. The first orca would locate the ray and then pin it to the bottom by the head, using the tip of its rostrum. This meant the orca was far enough away from the ray's tail for the barb not to reach. By continually swimming downwards the orca would prevent the ray from getting away. But in the meantime the ray would be flailing its tail around frantically, at which stage the second orca would come in and bite it – sometimes right off – effectively rendering the ray defenceless.

At other times the orca would spot a ray hiding under a rock ledge or asleep in the sand, with its tail sticking up to ward off potential predators. Unfortunately for the ray – at least where orca are concerned – its upright tail was anything but a warning. In fact, it was a strategically positioned handle. The orca would very carefully grab the tip of the ray's tail in its mouth and then gently lift it out of its hiding place. Hanging by its tail from the orca's jaw, the ray would remain motionless. At this stage the orca would follow one of three courses of action: rise slowly to the surface and flick the ray through the air like a frisbee; flip it upside down underwater; or let it drift off.

We know from studies of rays and sharks (collectively called elasmobranchs) that if you grab them and quickly flip them upside down they can be rendered helpless. They will lie there, motionless and defenceless, for minutes at a time. In scientific terminology this is referred to as 'tonic immobility', and it is possible that orca are flipping the rays to induce this condition and thereby reduce the risk of harm from their prey's defence mechanisms. Where the orca lets the ray drift off, as the ray starts to sink downwards it also seems to wake up and begin to swim off. Oftentimes the orca who has originally caught it now makes a move and grabs the ray and kills it, but at other times – especially if a young orca is around – I've seen a second orca immediately take the ray by the head and kill it with one quick bite to the brain. The ray is then shared by the orca, usually lying at the surface side by side as they tear it apart.

But rays aren't always defenceless, and late one cold, windy afternoon, near the end of 1998, I received a report of a dead orca floating out in the Hauraki Gulf – one of the America's Cup yachts out training for the upcoming regatta had seen it and contacted me. I was living too far away to reach the carcass before nightfall, so I alerted the Auckland Volunteer Coastguard which sent up their plane in a training exercise and quickly located it. Then they sent out a boat to secure the dead orca and towed it into harbour.

The next day a colleague of mine, Pádraig Duignan, a vet specialising in marine mammals, was able to come along with another vet, Jane Hunter, who was also a pathologist, to conduct a dissection. From sections taken of a tooth (just as with a tree, you could count the rings and work out how old the animal was), we established that the young female orca was between four and six years old, and from my catalogue I could ascertain that it was not an animal known to me. However, we found no food in the orca's stomach during the dissection and were nearing the end of the

It is very unusual to find a dead orca – typically, when they die they sink. This young female orca was found by a sailing yacht, and reported to me. The Coastguard helped to tow it back to shore, where we performed an autopsy. We discovered that it had died from an allergic reaction to stingray barbs, one of which was lodged in her throat. **Ingrid Visser**

procedure without having come across any conclusive evidence as to what may have contributed to her death.

Then Pádraig was cutting the skull off the remainder of the carcass in order to have a better look at it when we discovered a stingray spine lodged in the orca's neck. The spine was long – over fifteen centimetres in length – and stuck right inside its flesh. We then examined the rest of the skull and found another much smaller ray-spine lodged in the lower jaw, just below the surface of the skin. When we were cutting the vertebral column into sections, we discovered yet another, and this one was wedged underneath the vertebrae in the thoracic region (in the backbone, just above the lungs). From its position it must have entered the area after passing through the orca's stomach and puncturing the wall, before working its way to just below the vertebrae. This third ray-spine was also quite large (over nine centimetres long), and was enclosed in a capsule of hardened tissue, which suggested the animal had suffered an allergic reaction to the foreign object.

We came to the conclusion that the orca had either died of blood loss, which wasn't likely as I had observed a lot of blood coming from its mouth when we first got to the carcass, or that it had undergone an acute allergic reaction to the venom from the ray-spine which had caused its death. As it had been stung before (the spine wedged under the vertebrae had been there a while, as had the one in the lower jaw), this clearly wasn't the first time this orca had attempted to catch rays. Obviously, hunting for stingrays was not a risk-free method of obtaining food and evidently required a lot of skill as well as plenty of training sessions for young orca. I now viewed these practice sessions with a far greater appreciation of the perils involved.

As for electric rays, they have their own particular method of defence – a pair of kidney-shaped electric organs, one on each wing, that are used for protection and presumably to stun prey. The New Zealand electric ray is capable of delivering forty to fifty consecutive shocks and emitting a charge that can render a full-grown man unconscious for about an hour – in other words, these little guys (ranging in size from about 0.5 to 1 metre across) pack a real punch. Just how much of a threat this is to orca is unknown at present, but every time I've seen them with a ray, torpedo or otherwise, they either have them by the tail, presumably to avoid the sting or the electric zap, or upside down in their mouths, probably to induce tonic immobility.

New Zealand orca also feed on other potentially dangerous prey – sharks. I have found evidence of them feeding on six species: the blue shark, *Prionace glauca*; mako shark, *Isurus oxyrinchus*; school shark, *Galeorhinus galeus*; basking shark, *Cetorhinus maximus*; thresher shark, *Alopias vulpinus*; and smooth-hammerhead, *Sphyrna zygaena*.

While off Kaikoura one summer's day early in 1998 I watched a group of orca, which included Prop, hunting for blue sharks. As Prop wasn't very fast (nor presumably very mobile because of her injuries), she didn't get involved in the hunt and she always had a 'babysitter', an orca who stayed with her while the hunt was on. Two adult male orca, Danny and Bent Tip, were circling a shark. They seemed perfectly aware that this animal possessed the potential to hurt them, as they stayed outside of its bite range while they circled from below and around. One orca who was ready to move in was Ragged Top (the top of whose dorsal fin was completely missing, possibly bitten off by another orca, or a shark). As the orca circled faster and faster the shark stayed at the surface trying to keep an eye on all three. Suddenly Danny rushed towards it lifting his tail high out of the water – then turning as if spinning on his head he brought his tail down in a karate chop right onto the shark's back. Seconds later, Bent Tip rose to the surface and repeated the manoeuvre. Immediately, Ragged Top rushed from below and grabbed the shark, nearly biting it in half. Once the shark was dead, Prop and her babysitter Betsy came over to snack with the boys.

I am often asked if the orca can recognise my boat. As a scientist,

I have to say no – we have no evidence of that. But as a whale-hugger and someone who has spent many hours with the orca, I would say that I am certain they can. Another time in Kaikoura, while watching Prop and her group hunt for blue sharks, I had the most incredible encounter with them which would seem to bear this supposition out. They had located quite a large shark, and instead of Betsy or one of the other orca babysitting with Prop, Prop herself came over and stayed near my boat (I never got too close to a hunt, as I didn't want to affect the outcome by providing a hiding place for the prey). I felt extremely privileged to think that Prop, and the rest of her group, trusted my presence enough to let me take over the minding role on this occasion.

Watching orca in the Bay of Islands another day, I became an unwitting participant in a hunting party. Arriving on location after receiving a call from Jo Berghan, I immediately got ready to get into the water. Jo was calling out to me across the waves, but I couldn't quite catch what she was saying. I was just keen to get in and start taking photographs as the orca were milling around in one location and most likely were hunting or already feeding on something. Of course, I found out later that what Jo was desperately trying to tell me was that there was a shark in the water with the orca. As soon as I got below the surface I saw an orca called Miracle holding a mako shark in her mouth by the tip of its tail. This one looked huge (but in reality was only 1.2 to 1.5 metres long) and was swimming as hard as it could, struggling to get free. Suddenly Miracle let go and the shark took a few moments to realise that it could escape, then immediately started to look for a place to hide. Considering that the only nearby objects in the ocean were the orca, itself and me, it chose *me* to hide under.

Suddenly I realised just how big this shark was – big enough to take my leg off, even if it was only snack-size for an orca. A sobering thought, but not one that lasted for long, as Miracle immediately came rushing towards me and the shark – and she was moving fast.

The shark took off towards my boat, which by now had drifted closer, but trying to hide there was never going to work, and only moments later Miracle spooked it out, caught it and then proceeded to devour it with another orca.

Not surprisingly, I had only managed to take a few photographs, and had missed the critical one with the shark in Miracle's mouth, although I did get one showing both her and the shark in the same frame. As it turns out, this was the first time anyone had ever seen orca eating a mako shark anywhere in the world, so it was exciting stuff to watch as well as record scientifically – even if I was a little closer to the action than I might have planned.

The anticipation I experience when I head out to an orca sighting is not just over the behaviours which I am going to witness, but who the orca are. Sometimes I see the same orca two days in a row, but in different locations, or I might not see that particular orca for several years. The rate at which I find them varies from location to location. For instance, Whangarei Harbour is huge – about thirty-seven kilometres (twenty nautical miles) of channels exist and in some places it is more than five and a half kilometres (three nautical miles) across. This is a large area to search as the orca can be found in the channels or up on the sandbanks. When they are on the sandbanks they are usually much easier to find – especially in the case of adult males, as their dorsal fins jut right out of the water because it isn't deep enough for them to completely submerge.

However, when I get a call that there are orca in the open ocean, perhaps off an island, the chances of finding them are extremely slim. Orca don't necessarily travel in a straight line, and if they do, they might only do so for a few hours at a time, or even less. They might also change their direction for any number of reasons, such as suddenly hearing a group of dolphins off to the east when they had been travelling south. Therefore I use a 'directional' hydrophone to try to listen for their calls – this is a standard hydrophone which

I have placed in a funnel lined with soundproof foam, so that sounds coming in from the front of the funnel are louder than those coming from the sides or behind, meaning that the direction from which the calls are loudest is the direction in which the orca should be. Even so, it is really a hit-and-miss game, and more often than not I don't locate the orca.

But when I do, it is all worth it. One encounter I had with a group off the Bay of Islands happened on a calm sunny day in March 1996. Jo Berghan and I were again out on the water and had received a call that the dolphin-watching boat had found orca at Cape Brett. We headed out there, and sure enough, there was the dolphin-watching boat alongside some orca. Not wanting to disturb their encounter, we moved in closer to the coast where we'd spotted another dorsal fin in the distance.

As we approached the lone orca we both realised something was wrong. This was an adult male, but he was covered in parallel marks which were raised like welts. These marks were fresh, still pink, and in some cases had bits of skin hanging off them. They were also extensive, covering his face, back and dorsal fin. He was obviously injured, but the injuries didn't seem to be too deep, and the marks, although apparently bites, were more like tooth-rake marks – deep enough to cause permanent scars, but not going much deeper than the skin. He was heading in the same direction as the group, but closer in to the coast, and eventually the other orca headed over to join him. As they surrounded us there were many I instantly recognised, but although there were some features about the injured male I thought were familiar, I just couldn't place him even though we kept taking photographs until the orca headed off into the sunset.

I had seen photographs taken by Dolphin Encounter owner Dennis Buurman of another adult male orca off Kaikoura, known to me as Outlaw, with rake-marks similar to this. The marks were healed but still clearly visible, and they too covered the orca's face, back and dorsal fin. The adult male Jo and I had followed definitely wasn't Outlaw, who had a very distinctive fin with a number of kinks in it, although in this case the fin was certainly mangled and leaning to the animal's right. I didn't find out who he was until I got my slide film back a few days later and could make out the shape of the dorsal fin and the outline of the eye patch, and start sorting through my catalogue.

Based on the photographic evidence and some strong suspicions I'd begun to develop about this orca's identity, it only took a few glances for me to confirm just who it was – Top Notch – the adult male with a large notch out of the top of his fin. As his fin was now tilting over, the notch wasn't quite as visible, but when you knew what to look for, it was very apparent. I was fascinated. Top Notch had first been photographed in 1989, as an adult male, off Kaikoura. He was therefore at least twenty years old when I saw him off the Bay of Islands. Old enough to fight with other males for the right to mate – but was that what had been going on? No other orca researchers had reported orca fighting for mates. In fact, no real aggressive encounters had been witnessed anywhere until 1993, when, after more than twenty years of observation, Graeme Ellis saw a large group of PNW Resident orca harass a group of three Transient orca. The Transients were seen leaving the area with fresh tooth-rake marks, but as these orca hadn't been observed prior to the tussle, no positive conclusions could be drawn at the time as they could have received the bites beforehand.

Could Top Notch have been perceived as a threat to a young calf perhaps? No newborn calves were sighted during our hours with these orca and I hadn't ever seen orca from this group eating or chasing dolphins, only elasmobranchs and other fish. Besides, I had seen Top Notch babysitting young orca before, which made it seem unlikely that he would now be a baby-eater. The tooth-rake marks I had seen on him certainly looked like they came from other orca, but surely if an orca wanted to inflict more damage than that, it could. And I think that whatever happened to Top Notch also happened to the adult male Olav, the constant companion of the adult female A1. Although the top of A1's dorsal fin is missing, and has been from at least 1977, she probably lost it when she got tangled in fishing line. However, the top of Olav's fin went missing just a few years ago and I think he might have lost it in an aggressive encounter with other orca. Olav also now has multiple tooth-rake marks, such as those seen on Top Notch and Outlaw. As yet, I hadn't seen which orca were involved in these aggressive interactions, but I could only imagine that it would be a very full-on encounter should I ever get the chance to view one.

I have seen Olav twice since I first found him with the top of his dorsal fin missing, but no further sign has been seen of either Top Notch or Outlaw. I don't know if Top Notch had other internal injuries which may have caused his death, but the tooth-rake marks on Outlaw had healed and he certainly seemed to be in good health when he was last seen. It is hard to say, without spending every single moment with these animals, just what is going on in their lives. Perhaps Top Notch and Outlaw came across the same orca who attacked them last time, and came out second best this time round. Finding out one thing about an orca (such as new tooth-rake marks) frequently seems to pose more questions than it answers.

Certain locations have been identified as orca 'hot spots' in other research areas, but so far I've had no such luck finding a similar hang-out here in New Zealand. Overseas these tend to be areas where the orca like to 'groom' – such as Robson Bight, British Columbia. This is a beach with smooth pebbles and rocks which the orca rub on and they frequently visit during the summer months to massage themselves on the seafloor. Similarly at the Indian Subantarctic Crozet Islands, orca rub themselves on the long kelp fronds and during the peak season of sightings can be seen in one small bay almost daily.

Both these sites are located close to the area the orca feed in, but in New Zealand it appears that orca don't have the option of staying in one place longer than a day or so, before moving off to a new feeding location, and I have found that they are travelling big distances. My photo-ID tracking methods indicate that most are travelling an average of fifty to 100 kilometres per day around the New Zealand coastline – and this is without taking into account any deviations, chasing of prey, or just general meandering, as I know the orca do. This is a lot of distance to cover when you also consider that the animals have to catch their food as well as sleep during the twenty-four-hour period.

When hunting, New Zealand orca tend to hug the shoreline and follow underwater rocky reefs and check out secluded beaches. All of these are areas where rays are likely to be found, and given that this is the most common type of prey taken by our orca, it makes sense that their foraging strategies maximise their chances of catching this prey. The rays aren't stupid either, and even a few days after orca have passed through an area you still tend to find them in extremely shallow waters or hiding under wharves or rock ledges, which means it makes sense for the orca to move on to another hunting ground fairly promptly.

In the few instances where I have found orca in the same area two days in a row, they tend to be hunting in a slightly different habitat, even if close by. For instance, on the first day they might be hunting along the beach, and on the second inside the estuary or out in the deep channel nearby. Typically, though, the orca move off and head towards the next prime hunting area in pretty short order. And although they may be hunting as they travel, they tend to hunt, then travel, and then hunt again.

What I have also found is that some orca are only found in certain areas of the country. Miracle, for instance, has never been photographed off the South Island, but has been repeatedly seen off the north of the North Island. The group of four orca who ambush dolphins have never been seen off the North Island, and only ever photographed off Kaikoura on the east coast of the South. Yet there are also orca which are seen all over the country, like A1.

We can't be sure yet if orca are territorial, because by definition this would mean they would have to defend their territory, and I have never observed any behaviour resembling that – but then again, perhaps that is what the tooth-rake marks on adult male orca have resulted from. But we can say with some certainty that the areas in which an orca has been seen are part of its 'home range'. An animal's home range might have a core area in which it spends more time, and certain other areas which it might visit just occasionally as part of its 'life range'. It is possible that this is the case with orca, as sometimes I find them in areas where I haven't seen them in over ten years and in which I might never see them again. Just what are they doing there? It seems most likely that they are searching for food.

An adult male orca chasing a common dolphin. The large upright black 'sail' is the dorsal fin of the orca. The dolphin escaped, albeit with large tooth-rake marks on its side. **Ingrid Visser**

So far I have found twenty-seven different species of prey for the New Zealand orca. It is highly likely, based on the wide range of food sources identified to date, that there are more species to be added to the list, but recording this sort of data is time-consuming. First the orca have to be found, then followed while they are hunting. Then the prey has to be actually seen to be eaten (some orca are known to kill an animal and not eat it, so this species would not be added to the prey list), and on top of all that it has to be seen well enough to be recognised and identified right down to a species. I have seen orca eating fish, but have only been able to identify three types for sure: bluenose, *Hyperoglyphe antarchia*; leatherjacket, *Parika scaber*; and kahawai, *Arripis trutta*. Although I have seen them feeding on a tuna, I couldn't be sure which species it was. Fishers have reported them feeding on smaller fish, but again we couldn't get an identification, and one did see a New Zealand orca eating a sunfish, *Mola mola*, which are not uncommon around the warmer waters of this country, although their skin is poisonous and no one knows how orca get away with eating them.

As previously mentioned, some New Zealand orca also eat other cetaceans. I have seen them take dusky dolphins, *Lagenorhynchus obscurus*, and attack common and bottlenose dolphins, and have heard of them feeding on pilot whales, *Globicephala melas*; southern right whales, *Eubalaena australis*; humpbacks, *Megaptera novaeangliae*, and the endangered Hector's dolphin, *Cephalorhynchus hectori*. The last observation was from Les and Zoe Battersby, who at that time owned the Dolphin Watch Marlborough operation in Queen Charlotte Sound at the northern end of the South Island. They even managed to get a photograph of the orca carrying the dead dolphin in its mouth. The picture isn't close-up, but you can see the tiny snout of the dolphin and it can clearly be identified as a Hector's dolphin. Of course, such an observation raises all sorts of issues, such as what predation means for an endangered species. But in reality, it is something that is a natural part of both the orca and the Hector's dolphins' existence, so they must be left to their own devices.

Some other food which I know the New Zealand orca eat are penguins, salp (a type of jellyfish), and octopus. Orca have even occasionally been recorded taking squid in some locations around the world, although so far they haven't been recorded as prey here. On the other hand, pilot whales are known to eat squid, and one feature I have noticed at strandings of pilot whales is that many have their teeth worn right down to the gums. There is some speculation that the orca of the PNW Offshore population may be feeding on sharks because their teeth are worn down to the gums, with the theory being that the sandpaper-like shark skin wears them down. However, New Zealand orca eat a lot of sharks and I have yet to see orca here with the kind of tooth wear I've observed on pilot whales. Perhaps some chemical reaction with the ink of the squid accelerates the wear on the pilot whales' teeth, and something similar might be happening with the Offshore orca, although it is likely to be many years before we even begin to get any evidence, one way or the other, for such a theory.

Similarly, in almost fourteen years of looking at the prey of the New Zealand orca, I have never been able to find any hard evidence of orca here foraging on seals or sealions. Two species can be commonly found in New Zealand waters: the New Zealand fur seal, *Arctocephalus forsteri*, and the Hooker's or New Zealand sealion, *Phocarctos hookeri*. Both were hunted extensively, and brought to the brink of extinction, for their fur, which was in great demand in the 1800s. Their numbers have steadily increased since they were protected in 1946 when commercial sealing was prohibited. It is possible that New Zealand orca may have hunted for seals or sealions in the past, when numbers were greater, but switched prey types and just haven't switched back. The other possibility is that there once was a population of orca around New Zealand which specialised in hunting seals and sealions, but when their food source was killed off they died out as well. We will probably never know, unless some genetic markers discovered in the future allow us to tell what type of food an orca was eating and if we can collect enough samples from the few orca skeletons dotted around the country's museums.

There is some speculation by David Bain, a researcher based in the San Juan Islands of Washington State, that fish- and mammal-eating orca have developed slightly different lower jawbones. During a hunt this area of the skull is subject to damage by strong agile prey such as seals, so David has suggested that this area of bone has strengthened over many generations of orca which specialise in feeding on marine mammals (such as the PNW Transient orca). Perhaps a morphological feature like this will give us clues to identify whether New Zealand had its own population of mammal-eating orca. But there are other characteristics which could help us understand if the New Zealand orca ever did, or still do, hunt this particular prey, and these lie with the behaviour of seals and sealions themselves.

Where seals and sealions are hunted by orca they are particularly wary of ocean predators. From watching Transient orca hunting around seal haul-outs (areas where seals consistently go to get out on the rocks), Robin Baird has found that the seals are extremely vigilant, and if they spot an orca cruising along the coast they become hyper-alert and the message gets passed around the group very quickly. I observed a similar situation at Punta Norte in Argentina when the orca came in close to the beach to hunt for sealion pups. The pups seemed oblivious to the danger when they were alone, but the adults were extremely aware and their alert behaviour was passed swiftly from one animal to another along the beach. When a mother sealion wanted to take her pup over the reefs and past the patrolling orca, she was sure to take it well up on the beach. Thus, the seals in these areas have learned that orca spell danger.

On the other hand, it doesn't make sense to waste energy on avoiding animals that are not predators, or a predator who is not hunting. Robin Baird found that one Transient orca catalogued as

X10 had a collapsed dorsal fin and that the fin was slumped so far over it dragged in the water beside it. Overall, this orca looked nothing like a 'typical' orca, and sure enough the seals didn't take any notice of it and made no moves to avoid it, even though it was actively hunting. To compare an experience of my own, I watched dolphins off Kaikoura swimming in close proximity to the group of orca which are known to hunt for dolphins, so perhaps the orca were not giving off 'hunting' signals at the time. It may have been a case of the dolphins 'mobbing' or keeping an eye on the orca while they were not hunting, so that when they did start hunting, the dolphins would know where they were and wouldn't be caught in an ambush.

Furthermore, where there are two different types of orca, perhaps the seals and sealions are smart enough to figure out the differences between Transient orca who hunt them and Resident orca who eat fish. Robin Baird also found that the Transient orca start to exhale when below the water, thereby reducing the volume of their blow by the time they surface (and directly affecting the distance it can be heard over), and that they let their breath out over a longer period of time, again reducing the loudness of the blow. Are the Transients trying to outsmart the seals with this strategy?

In our home waters it turns out that both the seals and the sealions take no notice of the New Zealand orca. I have seen orca cruising past haul-outs, only about a body distance away from the resting seals – sure, the seals looked up, but then who wouldn't? However, even when the seals saw the orca in such close proximity, they didn't become actively alert nor did they warn their neighbours or young. Likewise, I've also seen orca swimming past and underneath both seals and sealions, and each time the potential prey just put their heads underwater and watched the orca swim by. They didn't try to escape or even take that much notice, lazily rolling onto their backs and watching my boat drive past next. So it seems likely that these seals and sealions haven't been on the menu for New Zealand orca for a long time, if ever.

# CHAPTER SEVEN
## Close Encounters of the Orca Kind

With all these differences in where various groups of orca go and what they eat, it would make sense to assume that they have individual personalities as well. To say such a thing is frowned upon in the scientific community, because it borders on anthropomorphism – attributing human characteristics to cetaceans in this case – but regardless of whether I should use the terminology, do orca display individual personality traits? Well, as you might expect from what I have just said, as a scientist I am obliged and strongly recommended to say no, orca do not have personalities. But as someone who works with these animals in their natural environment, who spends hours at a time watching them and has observed them in many different locations around the world, I would have to give an unequivocal YES. To me, they are as varied and idiosyncratic as the people I meet while doing my research.

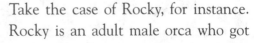

Take the case of Rocky, for instance. Rocky is an adult male orca who got his name because of the way he hunts – in among the rocks. Rocky has a hunting technique that is uniquely his own. He hunts for rays in shallow areas that are sometimes too small for me to wedge my boat into; he goes up onto beaches and chases rays along rocky headlands. One day I was out with a fellow researcher and computer guru, Terry Hardie (who I met during the second year of my study), and we were watching Rocky pass over a sandbar in Whangarei Harbour. Following closely behind, we could see he was chasing a ray, and I was keen to obtain a photograph of him as he chased this ray up onto the beach behind the sandbar. All of a sudden, the boat ran aground. We were in water so shallow that we could no longer follow Rocky, although he was steadily thrashing along ahead of us. However, we were in a bit of a predicament as the tide was falling.

I knew the boat, with its rigid-hulled aluminium bottom, wouldn't be damaged on the sandy seabed, but I wanted to keep following the orca. So I hollered at Terry to get out and start pushing while I stayed in the boat to help drive us off. Terry didn't have a wetsuit on, or a swimsuit, but he was wearing wet-weather gear and rubber boots. He didn't stop to take these off as I told him the tide was dropping quickly and unless he wanted to sit on that sandbar for the next six hours he had better get out and push fast! Terry didn't stop to check which side of the boat the sandbar was on and got out on the deep side, and was suddenly standing in water up to his chest. I was laughing so hard I couldn't even help him back into the boat once he had pushed (or rather pulled) us off the bar! I'm not sure Terry has ever forgiven me for that dunking, but during all of this carry-on Rocky continued to chase the ray and I never did get that photograph of him with it on the beach.

One aspect of studying the orca which has fascinated me, and which I think will continue to drive this project long after I have gone, is the interaction between orca and people. I have a theory about why

New Zealand orca are so interested in people and why they appear to go out of their way to interact with them, but it took a long time to form, and is perhaps best illustrated by some of the close encounters I've had with them over the years.

One day in September 1995, I was out in the Kaipara Harbour on the North Island's west coast following a group of sleeping orca. They had been feeding on rays on the shallow sandbars and it had been an exciting time watching them hunt, but now they were peacefully cruising up the harbour. It was a group without any adult male orca – not unheard of, but not common either. We were all travelling at about two knots (just over 3.5 kilometres per hour), and my engine was on idle as I ate my lunch and followed along behind. Suddenly the engine roared as the revs leapt to over 1000 rpm (but the boat didn't increase speed), and I quickly grabbed the throttle and took the boat out of gear, effectively stopping it.

Mystified, and not to mention just a little shaken out of my meditative mood, I looked all around to see what on earth could have caused such an event. I didn't appear to have anything tangled around the propeller and the engine revs had dropped back down to normal, so I put the boat back into gear and continued along. But it wasn't two minutes later when the same thing happened again, and once more I quickly took the boat out of gear and slowed down. I was starting to get a little concerned as I was a long way from the nearest boat ramp should something be seriously wrong with the engine. (For it to be inexplicably jumping up in revs, I thought there had to be something really wrong.) I decided to give it one more try, but this time I was going to closely watch the engine to see what happened.

Sure enough, not long after the boat was back in gear and I was following along behind the orca it happened again, but this time I saw what was going on (well, so I thought), as I noticed a huge burst of bubbles coming from the propeller. I figured there must definitely be something tangled around the prop, such as a plastic bag or some fishing line, so I stopped the boat again, turned off the engine and lifted it to have a look.

Going to the stern I saw a female orca swimming nearby, but given that I was surrounded by them I didn't really take that much notice. Checking and double-checking the engine, I still couldn't find anything wrong with it, so put it back down, restarted it, and began following the orca again. But then I saw the orca who had been travelling alongside heading in towards the stern of my boat. As she passed under the engine its revs suddenly leapt off the dial again.

Now I realised what was taking place, and quickly pulled the boat out of gear. The orca was passing under the propeller and blowing a huge cloud of bubbles into it, causing it to cavitate (spin fast through the air, instead of through the water) and the engine revs to spike, which led me to take the boat out of gear and stop it. Perhaps this is what she had planned all along, or perhaps it was just a game that developed, but it sure hadn't taken this orca long to come up with the equation: bubbles + cavitation = boat stopping.

When she came past the boat again I stuck my hand in the water and started wiggling my fingers. If she wanted to interact, I thought, why not offer some encouragement – and that seemed to be all the encouragement she needed. Moments later she was hanging next to my boat and looking right at me. As she lay beside the boat and started blowing more bubbles at the surface I put my face in the water and did the same thing. As I was 'talking' to her she could see my mouth moving a million miles an hour and perhaps she figured she should move hers. So she started opening and closing her mouth as she hung vertically next to the boat. I can honestly say that I never felt any sign of aggression from her, and even though she was opening and shutting her mouth right next to my hand, I didn't feel like she was attempting to grab or bite it. And just to give you an idea of the size of this wee lassie's dentures – an orca's mouth is about a metre across and they have ten to fourteen pairs of teeth in both the upper and lower jaw (that's forty to fifty-six teeth), all large, pointed, and protruding about five to ten centimetres from the gums.

Although the whole encounter only lasted about ten minutes it felt like ten hours. The boat had drifted quite a ways from the rest of the orca group and suddenly she decided that it was time to go and rejoin them. I followed along for about another hour, watching them sleeping, and going over and over the encounter again in my mind. Why had she blown bubbles into the spinning propeller in the first place, why was she opening and closing her mouth, and what significance did our interaction hold for her? All questions and definitely no answers.

Nine months later I was approaching another group of orca on the opposite coast, in the Bay of Islands. I was with Steve Whitehouse, who had been with me during my first encounter with Nicky at the beginning of my study. Less than a minute after we showed up a female orca turned away from the group and came 'submarining' across the surface – swimming half in and half out of the water. She was headed straight for the boat and it looked for all the world like she was going to play chicken and see who was going to turn away first. There was no way I was going to risk hurting her, so I immediately took the boat out of gear. Approaching the becalmed boat, she passed under it and then lay just below the surface where she blew a large burst of bubbles at us. Reminded of my encounter with the orca in the Kaipara Harbour, I stuck my hand in the water and wiggled my fingers at her.

The orca came up towards my fingers with her mouth open and lay beside the boat. Then she moved towards my hand and was coming in quite fast (or so it seemed!), giving me quite a fright, so I pulled it out. However, when I put my hand back into the water the orca came right up to it and slowly opened her mouth again. She seemed to realise that I hadn't been quite ready to interact and so she approached carefully and opened her mouth really slowly. As I moved my hand towards the front of the boat, she gradually moved forward with it. Then when I moved my hand out from the boat, she slowly moved out from under the stern, to lie right next to me. My face was less than thirty centimetres from hers and she looked me straight in the eye (her own eyes, while surprisingly small, had a huge black pupil). Then she slowly sank back below the surface and glided off, with her fin no more than five centimetres from my hand.

The whole encounter had left me flabbergasted (not to mention Steve), but my mind was working in overdrive. Although it was nine months previously and on the opposite side of the island... could it be – surely not? Shaking with excitement, I stood up and looked at Steve. 'It's her,' I said. 'It's her – the one from Kaipara!' Convinced it was my bubble-blowing friend, we drove slowly over to the group. The water was quite shallow (about 15 metres) and very clear. We could see the orca below the surface, moving along with the boat, and they slowly pulled ahead of us. When they surfaced as a group, one of them turned around 180 degrees and headed back towards us, again submarining. She was coming hard at the boat, so Steve took it out of gear and then turned off the motor. She slowly sank and the boat drifted over the top of her.

Once the boat had passed overhead she rolled over in the water, and swam up to the side where I was. Rising to the surface, she blew a really big bubble into my hand that was resting on the water. Then she lifted her blowhole out and blew a whistle-raspberry and came closer while rolling over and looking me right in the eye. She then looked at my hand, and I wiggled my fingers. At this point the orca pivoted, so her tail was hanging down lower and her snout was pointed towards my hand. From this angle I guess she had a better view of my hand at the water surface. She then slowly opened her mouth and rose in the water until the top of her snout actually *touched* my hand. She flinched, but didn't pull away, and rose again, to touch me once more.

Then she sank back into the water – almost as if to draw my hand beneath the surface. I obliged, slowly lowering my hand as she submerged, and she opened her mouth again and rose back towards my hand. I raised it as she came closer, and when it was just above the surface I turned to Steve, who was taking photos. He got a photo of my hand just above her open mouth as she rose a little higher and touched it with the tip of her top, open jaw. Next she slowly sank, rolled over, looked me in the eye one last time and swam off. I couldn't believe what had just happened. Here was a completely wild animal coming over to interact with a human. She had been offered no incentive such as food, yet had approached the boat twice and had just reached out to *touch me*! I knew that I would have to give this amazing orca a name, and later as I lay awake all night going over and over in my head what had happened, I realised I had the perfect name for her – I would call her Digit, after the gorilla studied by the ground-breaking researcher Dian Fossey in the late 1960s and early 1970s. Digit had been the first to reach out and touch a human. Afterwards I also found that Digit the orca was always with another female orca, whom I called Dian after Dian Fossey.

But that was later in the day, and for now I was still back on location and Digit had moved off with the other orca in the group. We were following along behind as they headed east along the coast. As we started to travel parallel with them Digit again turned and left the group and approached my boat. This time there was no hesitation from either of us as I put my hand out over the water (not in) and she rose up towards me with her mouth open. She nudged my hand with her bottom jaw, and I moved my fingers to touch her teeth. Her tongue was moving (quivering is the right word, I guess), and she opened and closed her mouth with my hand on her bottom jaw. Then she sank back into the water, rolled over, lifted her pectoral fin to within centimetres of my hand, and slowly glided away.

Digit, a New Zealand orca, lying next to my boat. She is very friendly and often approaches me with her mouth open, which other people might interpret as an aggressive move, but she is always slow and careful. I see it as one of the few ways in which she can show she wants to interact. You can just see her eye below the eye patch. **S. Whitehouse**

In all, we spent an hour with Digit's group before we had to head back to shore, but it was one of the most magical hours of my life. I still have trouble putting into thought, let alone words, just how awesome this encounter was. I also thought about what an incredible amount of trust Digit had shown me – to allow me to put my hand inside her mouth, to touch her teeth, to reach out and touch me. Although there haven't been any reports of orca killing people in the wild, we're still talking about an animal with the potential to kill a human and I had just put my hand inside its mouth.

And it was for this very reason that I received a lot of criticism about this encounter and subsequent similar ones. Some people claimed that my interactions with these animals were unscientific and that I would be unduly affecting the data I collected. And perhaps they are right, because I have seen things with the New Zealand orca that no one else has ever seen, and in many cases I continue to be the only one who sees them. However, when Dian Fossey and Jane Goodall started to habituate the primates they were working with (Jane had been working with chimpanzees), there was a lot of criticism of their work too. Opponents, in particular scientists, suggested that they weren't witnessing 'real' behaviour because the animals were influenced by the observers. But now these research methods are considered standard practice with primates, and although it may take years for them to become habituated, long-term research with groups who are so used to humans that they take no notice of them are yielding the most surprising results. Who knows, perhaps the unsympathetic cetacean scientists will see the value in such an approach in years to come as well.

Rightly so, though, some of the criticism centred around the idea that the general public, upon seeing me interacting with orca in this way, might try it themselves. And occasionally this has happened. People tell me that they have read about my work or seen photographs of me interacting with orca, and when they had seen orca themselves, they had given it a try. For some, the orca took no notice, but for others it seems the word is spreading, not just in our world but in the orca world as well. People have told me of young calves putting their heads on the back of boats, of adult male orca rolling over next to fishing boats to have their tummies massaged with high-pressure hoses, and orca following along beside horses running in the surf.

All in all, there seem to be more close encounters with orca reported in New Zealand waters than the rest of the world put together. However, I would like to make it very clear here that, given the orca's undisputed status as the sea's top predator, this sort of thing 'shouldn't be tried at home'. There is a fine line between interest and aggression with orca, and this can only be interpreted from years of experience with the animals.

There are a number of other orca who interact with me as well. One is a young calf I have known since I saw it on the day it was born – or very soon after, as it was so incredibly tiny, just like Kootie – and only just learning to swim. Its mother was known to me as Star (NZ47), and she gets her name from a star-shaped scar on her saddle patch. I had seen her in many different locations around New Zealand and spent hours observing her over the years. On this day in May 1997, I arrived to find the newborn calf bobbing around on the surface. As I watched, Star started suckling the calf and drifted over towards my boat.

The calf was extremely curious and came straight over to the boat, undulating at the surface like a limp rag. Not yet having the experience to navigate around objects in the water, the calf bumped into the boat, but Star was right there and slowly moved it away. Again it suckled and again it approached the boat. This time it tried to bite the side of the pontoon – but it couldn't get its little mouth wide enough to chew hard – which is just as well, because even as a newborn it had sharp, powerful teeth.

I carried on watching Star and her calf (who was number 102 in the catalogue, so became known as NZ102 and got the name 'Rinie', after a friend and research assistant Rinie van Meurs) for over five hours. Finally I had to head home when it was getting too dark to see them any more. But as time went by I had numerous other opportunities to see Star and Rinie (I still don't know if Rinie is a boy or a girl). Each time I saw them Rinie would come right over to the boat and try to involve me in some sort of play. One time Rinie brought some seaweed over to play with, and another time spent most of the day bow-riding at the front of my boat.

Another orca who instigates close encounters is Miracle (the female I talked about earlier, tackling the mako shark). She first came to my attention when she stranded on a remote beach near the very north of the North Island in 1993. She was only young, about eight to ten years old; I still don't know her age exactly. (I wasn't told about her stranding until the day after it happened, so I didn't get the chance to take accurate measurements which, when compared with the PNW orca, can give us an indication of age. Nobody took skin samples, so we didn't get a chance to do any genetic profiling of her either.) The volunteers on the beach worked hard all night to rescue her, and they told me afterwards that there were orca off the beach during the night, possibly waiting for the stranded animal, and when she was finally pushed off they all headed away.

No one got any photographs of the orca who were off the beach, but a few people took snapshots of Miracle. They weren't photo-ID standard photographs by any means, as they didn't show the dorsal fin or saddle patches clearly, but in one I could see a small black dot in her white eye patch. Four years later when I was sorting through a box of photographs trying to get any final matches before I conducted the analysis for my doctoral thesis, I came across this photo of Miracle again (she didn't have a name then, nor even a catalogue number, as I couldn't be sure she had survived past her rescue). The box of photos was labelled the 'GOK' box – the God Only Knows box – as I didn't know some critical data such as who the animals were, or when the photographs were taken.

I had just spent two weeks with Tanya Jones going through and double-checking my entire catalogue for mis- and missed matches and we were totally exhausted from looking at photographs of orca. But I just had a funny feeling that I knew the orca in the stranding photograph, so off we went again: on the hunt for a photograph which we thought we had seen, but couldn't be sure, in among thousands and thousands of photos, slides and negatives. After searching for two days we finally discovered an orca which I had catalogued as NZ63 (but with no name), who had a small rounded notch out of her dorsal fin and a tiny black dot in her left eye patch. We had a match! It was an absolute miracle that we found the picture, and an absolute miracle that I had managed to get a photograph of that one tiny little black spot, and another miracle that someone had taken a photograph when she was stranded that showed this same spot. Hence her name, Miracle.

That evening, as Tanya and I had a cold drink sitting out on the verandah, looking out to sea, we considered the implications of such a finding. Here we had an orca that had stranded and had been rescued off the beach. Now, four years later, we had a series of re-sightings that showed she was doing well and living the good life, catching rays and hanging out with other orca. This was phenomenal, but even more so when we considered the way that NZ63 had recently begun to act.

During one encounter I had with Digit there had been two groups of orca around, and one particular orca (NZ63) was very curious about what was going on and had been watching everything very closely. The next time I came across this orca (still only known to me then as NZ63), she had approached the boat and blown bubbles, exactly like Digit. NZ63 didn't reach out to touch me, but she sure was interested.

Not long after Tanya and I made our finding, I was out with the orca again and came across Miracle. This time, knowing that NZ63 was also Miracle, the very same orca who had stranded and been rescued, I took the boat over towards her. She came right over, and when I reached out she started opening and closing her mouth, just like Digit, and also like Digit she came and touched my hand. Again I was lucky enough to have aboard my friend Rinie van Meurs, who was also a photographer, and he managed to get some great photos of Miracle and me together.

It was about now that my tentative theory about the interactive behaviour of the New Zealand orca began to take shape. I wondered if the behaviour was somehow linked to strandings and subsequent rescues. When stuck on a beach the animals go through an incredible amount of stress, yet they are very aware of what is going on during a rescue and will even attempt to help by doing things such as lifting their tails when you dig below them. If they are that aware of people helping them, perhaps they are also aware enough to make that connection once safely back in the water? Perhaps Digit had also stranded at some time in the past, been rescued, and this was what started her interacting with humans out on the water. She might even have attempted to interact with people before, but maybe they had been scared of the 'killer whale' which was approaching them? Or possibly because the same person, in the same boat, kept turning up again and again to watch her, she took the first step? It is hard to say, and we will never know if Digit stranded, but as the numbers of interactions with the New Zealand orca population spread I can't help but wonder whether this is the trigger.

There was another orca youngster, called Squeak – after my sister Monique's nickname – who took to the interactions game with great zest. Monique was out with me in Auckland harbour one day in 2000 and we'd been following A1, Olav and a bunch of other orca (there were three groups travelling though the area, spread out over about five kilometres and hunting for rays). By now we had been with the orca for about three hours and they were starting to settle into a slow and sleepy travel mode, but one of the calves wasn't quite ready for a nap. This little orca came right to the back of the boat and started surfing in the stern wake. It was rolling upside down and then leaping right out of the water only centimetres away from Monique's face as she knelt near the stern. When she held her hand out over the water the calf would knock it like it was a bat. I had never heard so much squealing and squeaking and I wasn't sure who was doing the most, or doing it louder – the orca or Monique – so there was little doubt as to what this wee orca would be called. Squeak the orca spent nearly an hour playing with Squeak my sister until we were the ones who had to break off the encounter and head home. Whenever I call Monique she asks me if I have seen Squeak, who is still very vocal every time I see her, even though I have yet to figure out orca language.

So how exactly do orca communicate with us, if indeed they want to? They are capable of touch and sound, but do we really have any idea of what is going on in their minds? In reality, no, not at all. However, there are certain physical constraints that must be taken into account. For instance, orca have no hands or fingers to wiggle in the water like we do, although of course they do have pectoral fins and a tail. But if Digit was interacting and she moved her tail, it would result in her swimming away. If she moved her pectoral fins she wouldn't necessarily swim, but it would certainly cause her to pivot or roll away, and if she were to do this then perhaps she wouldn't have such a good angle from which to watch me through the water surface. So perhaps the most logical things for an orca to use are their rostrum, blowhole, eyes and mouth. And sure enough, this is just what Digit and the other orca were using – reaching out to touch with their rostrums, blowing bubbles and whistles and squeaks through their blowholes and producing other sounds, looking directly at me and following my movements with their eyes, and, of course, opening and closing their mouths. What is this behaviour, if not a form of inter-species communication?

The next logical interactive step with the orca seemed to be to spend time in the water, watching what they were up to. I have a special permit which allows me to enter the water with orca, as this is not normally allowed in New Zealand waters. I had been getting into the water with orca throughout my research project, but there were some encounters which were different. One was with Miracle and her mako shark, which I described earlier, and another happened with Star and Rinie. I was by myself, off the Tutukaka coast, and had been following the orca for a while. Not a lot was happening as they slowly headed south and started to go to sleep at around midday (orca will sleep at any time of day or night, just whenever the mood strikes them). I was keen to try and observe some sleeping behaviour underwater so started to get my snorkelling gear on.

I tend to snorkel with orca, rather than dive. Some underwater photographers claim that bubbles scare whales and dolphins, because they assume it signals aggression to the animals (some cetaceans are known to let out clouds of bubbles when startled or aggressive), but I have used scuba tanks around a wide range of cetaceans and never seen negative reactions. However, this isn't my reason for the lack of scuba gear — it's simply that trying to drive the boat into position and be ready to get into the water with all that gear on, getting in and out of the boat and then trying to keep up underwater with the orca when wearing all the dive gear, is just more difficult. So, generally, I use snorkel equipment.

I had the boat in gear and was slowly idling along next to the sleeping orca. As is often the case with all species of animals, the adults want to sleep and the kids just want to play. Sure enough, this was the case today, and Rinie came over to bow-ride off the front of my boat. I walked to the bow and hung over the side, dangling my hand in the water. Then I headed back and continued to get ready. When I had pulled slightly ahead and to the side of the orca, I took the boat out of gear and slipped over the side. I created a big cloud of bubbles as I sank down, and as soon as they cleared, Rinie was visible not even an arm's length away. I was completely surprised, expecting the orca to be about 100 metres behind me, but no, Rinie had kept up with the boat and was waiting for me as soon as I got in the water. No sooner did I roll over onto my side to start swimming towards the sleeping orca than Rinie grabbed my fins and started chewing and pulling on them. I stopped swimming and rolled over to look at my feet. Rinie was hanging just beside them, and if he/she could have been twiddling his/her thumbs and whistling nonchalantly as if to say, 'Who, me? No it wasn't me . . .' I swear it would have been happening. I was laughing so hard, water was flooding into my mask and I had to bolt for the surface.

After clearing my mask and taking a few deep breaths I rolled back over once more and attempted to swim towards the other orca, which by now were close by. Again Rinie grabbed my fins and this time I twisted slowly over into an upright position and took some photographs of them being nibbled. As the orca were now moving off, Rinie headed towards them and I flopped back into the boat, put it into gear and manoeuvred into position. I still had all my gear on, and was ready to get into the water again.

I always try and get into the water from the same position on the boat, so that my behaviour is a little more predictable for the orca; and as I got ready, there was Rinie, waiting at the stern, right where I normally slip over the side. In order to get in this time, I actually had to push Rinie to the side, otherwise I would have landed right on top of her/him. Rinie, of course, thought this was a great game and immediately took hold of one of my fins again and tried to swim backwards by 'sculling' – moving the pectoral fins in circles – but I still had one free fin and could swim with this, so off we went, with Rinie hanging on my fin like a dog pulling on a lead.

It certainly didn't make for fast swimming, and by the time I got into position for the photograph of the sleeping orca they had moved on again. Two more attempts at this and the orca were waking up and heading in for some more hunting, so it was high time for me to head home and get some dinner myself. On the journey back I was smiling so much I had sore cheeks, even though I didn't get any photos of the orca sleeping that day.

I had another wonderful underwater orca encounter, quite close to this location, in March 1999. I had two friends working with me as research assistants. One was Jen Schorr, who had participated in the wild orca research project I had worked on in Iceland and had been involved with the reintroduction of Keiko, the orca star of the *Free Willy* movie, to the wild. The other was Brad Tate, who had worked with spotted dolphins and humpback whales in the Caribbean. I had been telling Jen and Brad about my encounters with New Zealand orca for years and finally they had found the time to come and work with the project, and I asked permission for them to be included on my research permit. It turned out to be a bonus that they could both come at the same time because although neither had met the other before, they had skills which would complement each other when it came to working with me out in the field.

It was only their first day here, and as my father's house had a great view out to sea we went there to have a scout around for any cetaceans. We'd hardly been there ten minutes when I spotted some bottlenose dolphins. They were slowly moving south, so we decided to head out and have a look at who they were. I had been taking photographs of bottlenose dolphins along this bit of coast for years now, and sending them to Jo Berghan, who kept a catalogue. We had found from this that the dolphins she saw to the north and the ones researchers saw to the south, in Auckland Harbour, were often the same as the ones I saw off Tutukaka.

So, off Jen, Brad and I went. We launched the boat and were heading out of Tutukaka, as I figured the dolphins would be just to the north of the harbour. But I couldn't quite believe my eyes when I saw a fin right in the harbour entrance – they must have been travelling faster than I'd calculated. Then there was another fin and I realised my mistake – that was no bottlenose dolphin, but an orca. We had literally just stumbled across them. Meanwhile, a juvenile orca had come rushing over to the boat and started blowing bubbles – sure enough, it was Squeak. I put my face in the water and blew some bubbles back and Squeak rolled upside down and started mouthing my hand. Jen and Brad were amazed.

As we headed further away from the coast the orca stopped hunting for rays and began bow-riding and swimming upside down right beside the boat. It was time to get in the water and observe the social interactions between the animals. I had been watching one sub-adult male babysitting Squeak and I wanted to see what was going on underwater. I slipped in first and got Brad and Jen to follow me. I wanted them to try to get photos of the orca if they interacted with me, so that we could make some positive identifications. Only moments later we were completely surrounded by orca. They were all around us, under us, and – if we dived down below the surface – above us. As both Jen and Brad had extensive experience of swimming with large cetaceans, being surrounded like this didn't worry them in the least. The orca seemed to realise that they weren't afraid, as they moved in closer and closer.

As I was diving down to get a photo of the sub-adult male, Squeak came swimming up right next to me and then rubbed shoulders with me. It isn't easy to try to hold an underwater camera in front of you, keep swimming, compose a photo, and have a 1000-kilogram baby rubbing up against you. But I managed, and the next thing I knew the sub-adult male was lying at the surface watching me. I swam over to him and he just lay there and then opened his mouth and closed it. I put out my hand and began moving my fingers and he appeared fascinated and just watched while Jen took a photograph of us. Then I reached out and began rubbing his cheek. Jen swam up to us and I backed off. She started rubbing his cheek and I got a great photo of them together. Brad also came by, but underneath us, with three orca swimming all around him.

It was an incredible encounter and we just didn't know where to look next. There were orca all around us, rubbing up against us and one another. We were seeing games of what appeared like tag – one orca racing after the tail of another, until it caught up and nipped the other's tail, and then the tables being turned and the latter chasing the former. We saw orca rubbing their tails all along the underbelly of another and orca just lying at the surface taking in the whole thing. To be immersed in the midst of an orca group socialising like this was an awesome experience.

Suddenly all the orca turned away from us and pointed their heads in a southerly direction. What could be going on? Perhaps a shark, some dolphins, or something else? Whatever it was, the spell was broken and they moved off in a group. We were exhausted and swam back to the boat, and then realised what had unexpectedly cued the orca to turn and leave – not more than 100 metres away was a boat heading full steam right for us. It turned out they had spotted my research boat and the orca, and were coming over for a look. I was disappointed our encounter had prematurely broken off, but was pleased that someone else had had the chance to see these amazing animals.

The boat followed the orca for about ten minutes before turning back towards the coast and heading for home. We stayed with the orca a little longer, and only moments after the other vessel had left we were again surrounded. We didn't get back in the water as it was now time to head home ourselves, but it was a nice goodbye from the orca. I looked around at Brad and Jen who were just sitting speechless at the front of the boat and we smiled and turned for home and hot showers.

There was another underwater encounter with a friendly juvenile orca whose identity I never did discover. That orca was bow-riding on my boat one day, so I slipped into the water. It came over to check me out and appeared extremely interested in my camera in its housing.

In order to take high-quality underwater photographs, typically you require four things – a good camera, a good lens, good light, and your subject close. Because water is denser than air, if the orca or whatever you are photographing is too far away, there is a lot of water between you and the subject, and the picture looks like it is

This young orca could see itself in the reflection on the front of my underwater-camera housing and was bobbing its head up and down and twisting from side to side as it watched itself. **Ingrid Visser**

taken in a blue fog. But if you get too close to something and it is large, it's like taking a photograph of the side of a bus. Wide-angle lenses overcome the problem, but when using them underwater some extra issues with refraction are encountered. Basically what happens is that the camera is put into a waterproof housing, which by default has some air in it. The air between the lens and the front of the waterproof housing creates an illusion of a picture which the camera can't deal with, so the solution is to have a curved clear dome at the front of the housing. This corrects for the false image and allows the camera to work clearly underwater. What it also creates is a mirror on the outside of the dome.

The juvenile orca I had just got into the water with could see this mirror and could also see itself reflected in it. I hung motionless in the water, watching this little orca watching itself. It moved its head up and down, and so did the mirror image. It moved its head to the left and to the right, and so did the mirror image. During all of this I was floating there astounded – this was an unmistakeable example of self-awareness, something that had originally been defined as a characteristic only possessed by humans, and one which sets us apart from other primates like chimps and gorillas. In-depth studies of primates had discovered that they too had a degree of self-awareness, but I didn't know of any studies at that time which demonstrated that cetaceans also had it. Since this encounter, some such studies have been conducted on dolphins in captivity and they certainly exhibit self-awareness. Eventually I took a few photographs of the orca looking at itself in the mirror, just before it swam off to rejoin its group.

Suffice it to say that there has never been a time I went out with these animals that I wasn't amazed by them and that I didn't learn something new. I still remained the only woman working with orca in the Southern Hemisphere and the only person working with orca in the whole South Pacific and Australia and New Zealand region. I was the only orca researcher in the world who was getting in the water to watch these animals, and still, after working with them for over fourteen years, I continue to meet orca that blow my mind away. Do they have personalities? Well, I'll leave that up to you to decide.

# CHAPTER EIGHT
## Ben the Wayward Orca

One orca I have grown particularly fond of over the years is Ben (NZ101). I first encountered Ben when he was just a juvenile while watching him and the others in his group – Rocky, Spike, Double Notch, Venus and Miracle – hunting for rays all around the Northland coast. I had seen Ben playing with his babysitters and I had seen him attempting to catch rays. And he wasn't very good at it either. He was a typical juvenile – trying very hard but just not quite getting the actions right. The rays sometimes got away just by out-manoeuvring him, or by hiding under a rock or beneath a wharf where he couldn't quite reach them.

As Ben grew older he got much better at catching rays and acquired the skill of flicking them through the air like frisbees. On one occasion I watched Ben in Whangarei Harbour as he chased a ray over a sandbar which was just too shallow for him and he got stuck. By thrashing around and rolling on his side he quickly got himself free.

This was something which I had observed other orca do as well, and it is part of the high-risk foraging behaviour associated with chasing rays – an occupational hazard, if you like, that New Zealand orca simply learn to deal with. Little did I realise then that Ben wouldn't always be so adept at freeing himself, and that one day I would play a part in rescuing him from a sandy predicament.

We knew that orca did get stuck on beaches occasionally, and Miracle had been a prime example of that tendency. I had been involved with Project Jonah for a number of years and had been active in saving hundreds of cetaceans of a wide range of species, but up until that time I had never been to an orca stranding. Although New Zealand had, and continues to have, one of the highest rates of orca strandings in the world, I had either been out of the country or too far away to reach the animal before it was too late or before it was rescued.

So why do cetaceans strand? There seems to be no universal answer to this question, although they can strand either alive or dead. Dead stranded animals are usually referred to as 'beach-cast', to distinguish them from those that come ashore alive and then later die or are rescued. Strandings occur in three main types: single strandings; pairs, which are typically, but not always, mothers and calves; and mass strandings, which include groups of three or more animals. The mass strandings usually occur in highly social species, such as pilot whales, among which the lifelong social bonding is often so strong that if one animal strands, the others will follow, no matter how high the risk to themselves.

We still don't know for sure what causes most strandings, and some of the theories about what causes them are outdated, or just plain crazy, but others may hold the clues as to why these distressing events occur. Some of these bizarre and logical theories include:

- following ancient migration patterns that are now covered with land
- confusion from magnetic anomalies

- mass suicide
- injuries, including natural (e.g. shark bite) or man-induced (such as bullet wounds, net entanglement)
- sickness, including infections
- 'whale traps' – areas that naturally confuse echo-location, by absorbing or bouncing signals
- inefficient echo-location – i.e. whales can't 'see' shallow beaches, so just run aground
- old age
- population explosions – the more whales there are, the more will strand
- fear of drowning
- genuine mistakes
- lack of food – cetaceans strand to prevent dying of starvation
- trouble during pregnancy
- confusion during storms
- influence of lunar phases – e.g. full or new moons
- rejection from the social group
- parasites
- deep-water species finding themselves in shallow water and panicking
- pollution
- predation – orca, sharks or other predators chasing cetaceans onto the beach
- inner ear trauma (as well as other organs) caused by loud, man-made underwater sounds

and the list goes on.

But whatever the reason, at a mass stranding there is often one animal termed the 'key' whale. This is usually the individual that started the stranding, and is often the first up on the beach. The key whale may be the animal that is highest on the beach (as it may have stranded at the peak of the tide), or it may be the one that is furthest along the beach (because any current may have taken the other animals up or down the beach, before they too stranded). The key whale may even be in the centre of a group of animals, arranged like petals on a flower, and on some occasions may actually be buried under other whales who were so desperate to get it they swam over it when the water was higher, and are now piled on top of it. At some strandings there may be a primary key whale, with secondary key whales (who may be dominant females) distributed around it as well. It stands to reason that if the key whale is taken off the beach the other whales will try and follow it. And this forms the basis of the system Project Jonah and other whale groups use to try and re-float and rescue stranded cetaceans.

There are a number of factors that contribute to the success of rescuing stranded whales. These include early detection and notification. In New Zealand a team of dedicated whale savers is 'on call' for strandings, ready to mobilise at a moment's notice. Once located, stranded whales need to be kept wet and cool. The layer of blubber that keeps it warm in the water literally causes the whale to cook alive out in the open air, and is one of the primary reasons whales die on the beach. Covering the stranded animal with a sheet and keeping it wet helps to prevent rapid evaporation of the water, and in some locations helps to prevent sunburn. At some beachings the fire brigade is called in to help water the stranded cetaceans.

Another important factor contributing to the success of a rescue is getting the whale upright as soon as possible. In the ocean whales are nearly weightless, supported all around by water, but once stranded they have a fixed, and relatively strong, pull of gravity on one surface of their body. If they are left lying on one side they may become disorientated, and have trouble swimming in a straight line when released. Pools should be dug for the flippers and tail, and these should be kept filled with water. When first placed back in the sea and once they are floating, whales should be gently rocked from side to side, as this appears to help them regain their balance. Another reason to upright stranded whales as soon as possible is that any animals still in the water, but on their sides, may have extreme difficulties breathing as the blowhole may be submerged during breaths, and once stuck the animal cannot lift its head.

During a stranding many, if not all, of the whales are under considerable stress. It is essential that each whale gets individual attention, and they often form a bond with the humans looking after them. If a whale has a person tending to them for a few hours, and that person goes away, perhaps to get some sleep, the whale becomes very stressed; its heart and breathing rates go up, it lets out whistles and calls, and it will start trying to move around. When the person returns, the whale settles back down again. For this reason, it is now common policy to have three people assigned to one whale, working in overlapping shifts until the animal is successfully refloated. This human contact is very important, and whales can be seen to observe the people working around them.

If you should ever come across stranded cetaceans there are four main things you should do:

1. Notify the appropriate people, e.g. DOC or Project Jonah (in other countries this is often the police).

2. Keep the whales cool and calm. Cover them with sheets if you can, and pour water on them. Prevent barking dogs from approaching, as they upset stranded whales.

3. Get the whales upright (this should only be done by people who know what they are doing, as whales can easily be injured – fins snapped off, shoulders broken).

4. Do NOT return the whales to the water – wait for people who know what to do. If you return a whale to the water when it is not ready to go, it may re-strand at another location and die.

More and more science is being conducted while the animals are still alive on the beach or after they die. Research on live animals has included the taking of measurements, sexing individuals, and collecting small skin samples for DNA analysis, all of which are used to determine the social structure of the groups. Calls, whistles, and echo-location clicks are recorded while the animals are on the beach to provide clues to their use by the whales. These noises could possibly be used in the design of 'whale-proof' fishing nets. In the water, research has also included recording sounds, as the animals are very vocal once reunited with their group. Interference with the cetaceans is avoided when they are trying to re-establish social bonds as it is thought that this may cause another stranding.

When stranded cetaceans die, they are not lost to science. The carcasses are often used to try to establish the cause of the stranding. Individuals are cut open to discover if they are sick, or have parasites. In the case of many beaked whales the throat is often found to be blocked with plastic bags (these whales are thought to eat jellyfish and may mistake the bag for one). Stranded female key whales are often found to be pregnant, and in some cases, where the full-term foetus is around the wrong way, this may have been one of the factors causing the stranding.

Over the years, some scientists have considered that stranded whales could not be saved – that the pressure of sitting on the beach would be enough to crush their internal organs causing death, that the stress levels from a stranding were so high that a whale would die even if it was put back into the water, and that stranded whales would just try and re-strand and continue to do so until they died.

However, I've been involved in many necropsies (animal autopsies) of dead stranded cetaceans over the years, and I've never seen crushed internal organs. Besides, many whales dive to tremendous depths without any ill consequences, and the pressure at these depths is far greater than anything they could experience on land.

In New Zealand, where an average of one stranding (of all species of cetaceans) occurs every week, until recently no hard evidence was available to show that whales that were re-floated did in fact survive. Photo-IDs are now taken of each cetacean at New Zealand strandings, and we tie cotton tape around the tails of live animals, so that when they are released the tape identifies them as animals that were involved in the rescue. Should they come ashore again within a few days, either alive or dead, we can tell that an attempt had been made to save them. Generally, most cetaceans, if a rescue is conducted efficiently, will not re-strand, and very few carcasses wash ashore. And this is where Ben comes back into the story.

In June 1997 I had been watching Ben hunting for rays in Whangarei Harbour with the rest of his group. At the end of the day I headed home wondering when I would see him next, but not realising it was to be so soon. The very next day I got a call that there was an orca stranded on a beach about twenty kilometres to the south of the harbour. It was a shallow, sandy beach, and I suspected he had been hunting for rays and got stuck. Help was called for, and his rescue began. He was uprighted and a pool of water formed around him, to keep him wet and cool. Later, as the tide dropped, he was shifted higher up the beach, so that he wouldn't be injured in the surf during the night when the tide came back in.

When I had initially been called I'd been told that the orca was bleeding from the mouth and was going to be shot. As I was a long distance from the stranding location and wouldn't have got there in time to stop this going ahead, I made the decision to charter a helicopter to reach the orca before the guy with the gun did. It was going to be an expensive exercise, but as it turned out, the pilot,

Graeme Webber, ended up sponsoring most of the cost, as he thought saving an orca's life was worth it too.

When I arrived, the orca had been uprighted and appeared to be uninjured externally, except for a few minor wounds consisting of several fresh and superficial tooth-rake marks on his body, but these couldn't have caused all the blood which had been seen. He also had a cut running along the joint of the left pectoral fin and two blisters on the same joint – these wounds could have been caused when he rolled around in the surf. Another larger blister (approximately twenty centimetres long, five centimetres wide and five centimetres high), just below the base of the dorsal fin on his left side, appeared to have been caused during the stranding, by the fin drooping over at a fifteen-degree angle causing pressure on this area.

Recognising Ben immediately, there was no way I was going to let him be shot. The blood had stopped, and although some people thought that it had been coming from his mouth, when I gave him some salt water to drink, I could see that his tongue and teeth all looked fine. I set about collecting data, including a small sample of loose skin from one of the cuts, for genetic analysis. Dorsal fin, saddle and eye patch photographs were taken, and body measurements made, and from these we found out that Ben was nearly five metres long.

We administered basic first aid to him, by keeping him cool with wet sheets and bucketed water (this was later supplied by hoses from a fire truck), and dug holes in the sand to allow his pectoral fins to hang at a more natural angle. When orca are free-swimming their pectoral fins hang downwards, and for Ben to be lying flat on the beach, with his pectoral fins sticking straight out from his sides, was no doubt very painful, especially as one of the joints was damaged (and possibly broken) from rolling around in the surf. Adult male orca have large pectoral fins; Ben, who was only a teenager, had small pectoral fins, but they were large enough to dig big holes for.

As we dug the first hole, Ben was watching us closely. He was whistling and chirping away, and I was recording the sounds. I had also asked some of the volunteers to keep a record of how often he was breathing. If the rate increased, he was likely to be in stress, but if he held his breath for too long something might also be wrong. We had to dig deep enough for the fin to hang downwards, but not so much that he would fall into the hole. Once we had finished digging on the left side, Ben gave a huge sigh of relief and closed his left eye. As we walked around to the other side to dig out the other pectoral fin, Ben rolled himself over as much as he could, to allow us to get underneath the right fin to start digging. He was very aware of what was going on and still had his eye open on the right side so as to watch what we were doing. We also dug a pool for his tail flukes and filled it with water, to assist in keeping his body temperature down.

The next morning, before dawn, we began to prepare Ben for his release back into the water. We could see in the daylight that his left pectoral fin joint was most likely broken, since it was hanging at a different angle from the right. The joint was also bleeding slightly from the cut running parallel to the body. However, as we didn't have anywhere that we could take Ben for him to recover, and as I didn't want him put into a captive situation where he might not be released, the decision was made to put him back into the sea, even though he was injured. I figured that although he might survive with a broken joint, he wouldn't survive with a broken heart if he was put in a tank for the rest of his life.

By 10 a.m. the tide had come in far enough for us to get him out through the surf in special rescue pontoons. Again, he seemed to know exactly what was going on, because as soon as he had water lapping around him he started to use his tail to help us move him. My computer guru friend Terry Hardie got an amazing photo of us all straining on the ropes and pulling on the pontoons, with Ben holding his tail high in an effort to help. We used the local surf-rescue boat to help tow him out through the surf. As soon as he was in the water he started calling louder and louder, possibly for the rest of the orca in his group.

Once out in deeper water, we deflated the pontoons and removed them and the PVC/nylon mat which had been carefully slid under his belly back on shore. Ben, probably because he had been lying on his left side when he first stranded, rolled to the left, got a little tangled in the mat and started to thrash around. As I was wearing my snorkelling gear I swam up to him and managed to make eye contact. As soon as I did this he lay still, even though he was still tangled up. I had to get the mat over his head and off his blowhole so he could breathe, and the only way to do that was to lean against him and lift upwards. However, Ben's skin was slippery, so I ended up putting my hand inside his mouth and wrapping my fingers around his teeth so I could get a grip. Not a very safe thing to do, in retrospect, but desperate means call for desperate measures, because if we hadn't removed the mat he would have suffocated. Ben took it all in his stride though, and as we were removing the mat he started to flex both pectoral fins in slow circles, suggesting perhaps that the injury to the left joint was not substantial.

As we finally released him he rolled over and looked back over his shoulder right at me. It was the most remarkable thing – he knew that we had been helping him on the beach, and now he seemed to be giving us one last thank-you look before heading off. We were ecstatic, but soon afterwards the adrenalin rush started to wear off and although we watched Ben for a short while as he headed north, we had to return to shore. However, some locals who had been looking out to sea from their clifftop house during our rescue efforts had noted a single unidentified orca swimming parallel with the beach about seven kilometres offshore. They saw that when we placed Ben in the water the offshore orca turned and headed towards the coast. An hour after release, these same folks saw Ben join up with the unidentified orca, and both head north towards the Hen and Chicken Islands.

The very next day, Ben was re-sighted at the Hen and Chickens with Miracle, Rocky and the others in his group. I was too exhausted to head out there, but a film crew got some great underwater footage of him swimming with the group. Apparently he came close to the cameraman a number of times, quite possibly trying to see if this individual had been involved in his rescue the previous day. It was extremely satisfying to know that Ben was doing okay after his ordeal and that he was back where he belonged.

And nearly four months later I saw him again. He now had a big white scar on the left side of the base of his dorsal fin, a result of the large blister he'd got when his fin had drooped over on the beach. Fortunately, this blister didn't seem to have caused him any long-term harm, and the resulting scar was now an easily identifiable mark, even from a distance. In fact, it has turned out to be so identifiable that one particular day – 21 October 1997 – when some orca turned up in Auckland Harbour and a TV news crew asked me to fly with them in their helicopter to look for them, I could spot Ben from the air and tell them who it was. I'm not sure they believed me, but I certainly had no doubts about his identity.

You would think that Ben might have learned a valuable life lesson from his lucky escape – but no, he was young, and like many teenage boys just didn't seem to be able to stay out of trouble. It had been over a year since Ben had stranded and I hadn't seen him, nor the rest of the orca he travelled with, for a while. Then I got a call that there were orca sighted in the Bay of Islands, so off I headed.

Catching up with them near Cape Brett, I was dismayed to see an orca which had its dorsal fin split right in half. The split started at the tip of the fin and went downwards to the base, cutting cruelly into the animal's body – I couldn't believe my eyes. At first I thought the orca must have got tangled in a net or line and had injured itself as it tried to struggle free, but upon closer inspection it was obvious the cut was too clean for that. There were also three deep parallel gouges further forward on its back – clearly it had been struck by a boat propeller.

As we cruised alongside I tried to take as many photographs as I could. I wanted to know who this orca was and to be able to show people the damage caused when boats and orca collide. Moving the boat to the animal's other side, I gasped in horror. There was a large white blister scar below the fin – it was Ben. Had he, like Miracle, wanted to interact with people so much after his stranding and rescue that he'd approached too close to a boat? The wound was obviously fresh, and whoever had been responsible for the hit-and-run had to have known at the time that they'd collided with something. Unfortunately, despite pleas I put out in the media to try and find out what had happened, no one ever did have the courage to own up to having hit him.

Ben, meanwhile, was travelling very slowly. Getting into the water with him to try to have a good look at the injury, I could see that the back half of the fin was wobbling from the water pressure against it as he swam forward. When he surfaced, the back half of the fin also leaned over to the left – the same side towards which it had leaned when he stranded. With all this movement the base of the cut was continually being forced open and was ripping further and further apart. Ben was literally being torn in two. There was blood coming from the wound and I just didn't know how he could survive. Unfortunately I was running out of fuel so had to head back to port without knowing if I would ever see Ben again – and if I did, whether he would be alive or washed up dead on a beach somewhere.

A month later I was again called out to the Bay of Islands and Ben was back as well, but not looking good. The wound on his dorsal fin was now all infected and swollen. The tissue was all granulated and covered in large knobs where the water had been forced into the flesh. There were rotten pieces hanging off the edge of the cut and pus was oozing out the bottom of the wound. It was ghastly, and I should imagine extremely painful. If I had given Ben very little chance of survival a month ago, I gave him almost none now. He didn't look skinny, but because of their thick layer of blubber it can take orca a good couple of months to begin to look thin.

The back half of Ben's dorsal fin was still under pressure from the water as he moved forward. But he couldn't just give up and stop swimming – he had to keep up with the rest of his group and keep hunting for food. It was possible that the other orca were providing him with some food, but in his condition it wouldn't have taken much for a shark to finish him off, so he needed his group primarily for protection. As he swam forward, the rear half of the fin would buckle and flex and start to fold backwards. All this was keeping the base of the wound open and continuing to split Ben in half. It was very distressing to see.

I wondered if there was some way of getting antibiotics into him. But how much would you give him, and even if I could find some immediately, would I be able to get out and find him again in a hurry? Besides, even if it were feasible, I only had two days in which to accomplish it, as I was heading off to Antarctica for a whole season of work.

As I drove home that night I thought about it more, and contacted some vet friends of mine. They explained that we would have to ensure that the antibiotics were administered into the muscle and not the blubber, so there was no way of firing them from a gun; instead they would have to be administered by a carefully placed injection or at the very least through his mouth via food. I knew that Ben wouldn't take food from me, even when he was this sick, and there was no way we'd be able to arrange DOC permits to capture him to give him the antibiotics. So I headed off to Antarctica with a very heavy heart. I suspected that Ben couldn't survive such a traumatic wound, and was deeply saddened that humans had caused it, especially after they had also done so much to help him during his stranding.

Four months later I came back from Antarctica and carried on with my research. By now I had resigned myself to the fact that Ben must have died out at sea, as nothing had been heard of him and while I was away no orca deaths had been reported either. Then, one day, nearly nine months after he had first been injured, I got a call that there were orca to the north of Auckland. I headed off and finally found them, not long before dark, but it was still light enough to see one very special orca – Ben. It was definitely him, and incredibly his dorsal fin was looking all right. It was still split, and the back half had collapsed completely and was dragging in the water beside him, but the edges of the wound had healed up fine. There was clean, healthy skin covering nearly all of it, and although it was still slightly open at the base, there was only a little bit of pus coming out.

He came directly over to my boat and travelled beside me as I photographed him. I felt tears welling up in my eyes as I thought about the agony this poor animal had been through and how he still had the courage (or was that naivety?) to approach human beings. Whether he recognised my boat or not, I was grateful that he was still alive and appeared to be doing well. He seemed to want to show me just how well as he began surfing in the boat's wake and playing with a juvenile in the group. I watched, delighted that there really was a happy ending to this part of his story.

I have seen Ben on a number of occasions since and he is still doing fine. He doesn't seem to hunt as often as he used to, but that may be because he is now a prime contender for babysitting. Nearly every time I see him he is with a juvenile or a calf, and if he catches a ray he always shares it with them. He still comes over to my boat but I have definitely seen him avoiding other vessels. Perhaps Ben the wayward orca truly *has* learned his lesson?

There is also a happy outcome to Miracle's stranding story. In 2001 Miracle had her first calf – Magic – irrefutable evidence that stranded cetaceans can still be viable, i.e. produce offspring. A number of people have argued that stranded cetaceans become 'damaged' during a stranding; so that even if they survive they can't breed, and thereby become a drain on the population, and so should be killed when they are on the beach. The re-sighting of Miracle with a calf and Ben as a babysitter is proof that rescuing stranded whales can work, and that the animals have a healthy chance of surviving for months or even years after rescue and of making a valuable contribution to the population.

DOC and Project Jonah whale-rescue teams have the highest success rate for refloating whales in the world (ninety per cent as I write this), and it just goes to show that cetaceans aren't doomed the minute they strand. With a colleague of mine, Dagmar Fertl, based in the USA, I published my experiences with Ben in a scientific paper. I wanted researchers and rescue teams to know that their work was definitely worth the effort (however, at that stage Miracle hadn't yet had her calf, so I couldn't include that significant information). The paper seemed to have the desired effect, as a few years later, while I was at a conference, I had one researcher come and tell me how he'd successfully used my arguments to allow a rescue of a stranded orca to be attempted when authorities had been going to kill it. He said that he had since seen the orca alive and well and that he wouldn't have even tried to save it if he hadn't read about Ben's marvellous reprieve.

# CHAPTER NINE
## Filming Orca in the Wild

Underwater film-maker Andrew Penniket and I met while we were working on an eco-tourism ship in the New Zealand Subantarctic Islands. I gave a lecture about my research and Andrew asked me if I thought we could film the orca. I was sure we could, so we set about planning to make a documentary. As I was still carrying out my doctoral research at the time, a film project seemed a logical way to fund the research – the documentary makers would pay for all expenses while we were filming as well as reimburse me for my assistance.

Wildlife film crews usually spend many weeks, if not years, in the field trying to film animals and often they have to sit around waiting for interesting behaviour to happen. This film was going to be no different, and in the end proved to involve more waiting around than we had ever anticipated. The film crew varied in size from just Andrew and a soundman, to Andrew, a soundman, a production assistant, a sound assistant, an underwater cameraman assistant, an assistant producer, and sometimes their wives or kids. Along with all these people, I also had my own research assistant working with me, Danielle Mayson (known to all of us as Danny). This was a lot of people and equipment to try to fit into my tiny little research boat, and all in all it just wasn't going to work.

Andrew and I decided that we needed a second boat, and eventually swung a deal with Naiad, the makers of my own boat. We got a good price on a new boat, and they got good coverage in the documentary. The quiet four-stroke engines were supplied under the same deal with Yamaha, and the Lion Foundation provided funding for a trailer. Suddenly we were a two-boat team. As all this new gear had been offered at a good rate our supporters wanted to see me driving around in the documentary in the 'new' version, so I got to have the new, bigger boat and Andrew and the rest of the team squashed themselves into the tiny one. However, when the cameras weren't trained on me, most of the crew were with me in the bigger boat.

We travelled all over New Zealand looking for orca. I set up a toll-free number for people to call if they spotted orca – 0800 SEE ORCA – and we had over 50,000 stickers printed to distribute (I have since seen them on the backs of vehicles, on the sides of boats, at boat ramps and club notice boards, to mention just a few places). During the filming I was still giving talks about my research and requesting the public to call as soon as they saw orca, so every time I gave a talk I also handed out stickers. And as we drove around the country we popped stickers into thousands of letterboxes, with a little flier telling them about the research project and the film we were making. Even now, eight years after making the documentary, I still get people telling me that they have those stickers and fliers on their refrigerators – and I expect to keep my freephone number up and running until the day I die!

One of the main areas in which we were hoping to find orca was the Northland coast. Based on my database of orca sightings from previous years I had made plots by month and by location, and I knew the best time to see orca in Northland would be during the winter months. So the film crew set up house in a local hotel and we waited. It wasn't too long before we got our first call, but it took the crew over two hours to get ready and out on the water. Needless to say, I wasn't impressed, and even less so when we didn't find the animals. There and then I decided that I should just carry on as if they weren't there and they could tag along when they made it. This generally worked well, and when I got a call about orca I would phone them, give them all the details I'd received, and set off as fast as I could.

Typically, I was leaving within fifteen minutes of getting a call and would sometimes be with the orca even before the film crew had launched their boat. But at other times, when they were closer to the launch site, they would arrive first and could film the reaction of the orca as my boat arrived. Although I had switched boats it didn't take the orca very long to work it out, and soon enough they were treating my new boat like an old acquaintance. Digit was blowing bubbles at it, Miracle was coming right up alongside, and Ben was surfing in the wake (at that stage he had not yet stranded, let alone been hit by a boat).

When I had taken Professor John Craig out on the water with me he had been astonished at how quickly I could identify individuals from the boat. He also remarked on how I had amassed enough data to be able to predict where the orca were going to be, to some degree, and when out with them to predict their behaviour. Being able to do this is the key to filming any type of animal – that is, being able to anticipate what is going to happen so that the film crew is ready and in the best possible place to shoot. There were plenty of occasions on which I told the crew what was about to happen, and in many cases even which orca were going to do what. I knew, for instance, that Rocky would typically pass through narrow gaps in the rocks while hunting for rays, and that when he reached a headland he would usually wait there for Venus and the rest of the group to catch up.

Armed with this knowledge I was able to help the film crew set up in the most desirable locations, bearing in mind requirements like lighting and background, to film our cetacean subjects. But it wasn't always easy. There were plenty of times when the wind was blowing and the spray was coming over the side of the boat, making it extremely difficult for the cameraman to keep his gear dry. Given that this equipment was of such high quality that it was worth more than my boat, it was an important factor to consider. Then there were times when the orca just wouldn't do as we had hoped, and instead of staying and hunting off a beautiful beach with sheep dotted on the hills in the background, they headed out to sea and had a snooze. There were also plenty of times when I had to wear the same clothes each and every occasion we went out to look for orca, for months at a time, because we needed a linking sequence between two shots, and that would require me having my hair tied the same way and having on exactly the same shirt or jacket.

What we really wanted to film was orca hunting. No one had ever filmed them hunting for rays or sharks before and we thought we could do it. We needed perfect conditions though – an area with a lot of rays, clear water so we could see what was going on, good sunlight to reach down through the water, and of course the orca hunting – not to mention us actually being in the same place as all of this. It was a logistical nightmare, and Andrew had the film company executives breathing down his neck, impatient to see the 'rushes' – the uncut film that had already been shot – but as yet we didn't have any footage of the orca with the rays. We had orca swimming and orca surfacing, and Rob Brown, the cameraman, had even managed to capture one of the incredibly rare New Zealand orca breaches – where the animal jumps right out of the water – but as yet no orca with rays. The pressure was on. Every time the phone rang my heart pounded and I prayed it was orca, close at hand and willing to cooperate.

Finally we got a call that there were orca in Whangarei Harbour. This wasn't the clearest water we could wish for, but given that most of it was shallow, there was a chance it would work. So off we all raced. We got out on the water to find that it was Rocky, Spike, Venus and Ben, and we watched them chasing rays for several hours during which Andrew got in the water a number of times and caught some incredible footage of the orca holding rays in their mouths and food-sharing underwater. It was a dream come true.

At the end of the day we all headed off home very content with the day's work. Back at the hotel one of the new assistants took the film out of the camera and put it into the video machine to take a look at what Andrew had filmed. The machine, one of those cheap hotel versions, duly started to 'eat' the tape. Rather than stop it and try and salvage what was left, the panicking assistant pressed fast-forward – the entire tape was destroyed in moments. What a disaster! Months of waiting and months of following orca had just disappeared in a flash. I only heard about this some days later as the film crew didn't have the heart to tell me at the time. We never did get another chance to film orca food-sharing a ray either.

However, we did get to witness food-sharing with a different prey, and in another location. This time it was off the Three Kings group of islands, about a day's sail north of New Zealand. We were on a charter boat and I had heard that the orca were taking fish off longlines in the area. Sure enough, on the day we turned up the orca were there, taking bluenose off the lines as the fishing boats hauled them in. To try and distract them, one of the guys had put a fish on the end of a long rope and was trailing it behind the boat and 'fishing' for orca while the real work was being done alongside. The orca looked to be enjoying this game, as they played around with the fish for quite a while.

Figuring we might give it a go as well, we got a fish from the fishers, put it on a rope and started hauling it along behind us. Sure enough, a juvenile orca came straight over and started swimming along just behind the trailing fish, but upside down. As we pulled the rope in, it came in as well, and when we let the rope out, it hung further back. A friend of mine, Cheli Larsen, who was along to help out, had the rope. The orca calf suddenly took the fish and the line in its teeth and began pulling. It was pulling so hard that Cheli was nearly yanked over the side, but then the line went slack and Cheli lurched backwards.

The baby orca had the fish but obviously didn't want to finish playing. It swam up to the back of the boat and lifted its head out of the water, still with the fish in its mouth. Cheli leaned right over and took the offering. The calf swam alongside, watching us, and was then joined by an adult female, presumably its mother. The female swum right up to Cheli, rolled over, looked at her standing there with the fish in her hands and, as if to say, 'Give that fish back to my child!' rolled back over and started blowing raspberries at Cheli. To keep everyone happy, Cheli leaned back over the side and returned the fish to the calf who then swam off and shared it with another orca.

The water around the Three Kings is renowned for its clarity and also for its fish life. This was why the fishers, and presumably the orca, were there. However, they weren't the only predators in the area, as a number of people had seen and caught large sharks around the islands, so it was with some trepidation that I got into the water the next day to see what the orca were doing when they were taking the fish off the longlines. Was just one orca removing the fish and then supplying the others with the bounty, or were they all attempting to remove them? As I slipped into the water next to the fishing boat, I was keenly aware that there were only three of us in the sea – myself, the cameraman, and Cheli, who was there to watch our backs. As the currents were pretty strong in the area, we had underwater scooters. Predictably, the orca were very curious and came right over to us, hovering near enough for me to take good underwater identification photographs and for the cameraman to obtain some excellent footage. We watched avidly as each of the five orca took fish off the longline, and then shared them with one another. For me, one of the most fascinating things was getting the opportunity to watch a calf suckling from its mother.

We had seen so much and were so busy filming everything that, before we knew it, we had run out of film. We were a long way from the nearest location we could buy any, and we didn't want to miss the opportunity to witness the orca feeding. Fortunately, we managed to get a message to a friend, Grant Harnish, who owned a small plane, and he got hold of the TV company. They got some film to him which he packaged up in a waterproof drum and then flew out over the islands. He dumped the drum over the side, we steamed over and picked it up, and we were able to continue filming for another two days – and got some more incredible footage of the orca taking the fish off the longlines.

That wasn't the only time we used planes during the filming of the documentary. Although I had anticipated that we might be able to conduct a complete aerial survey of the New Zealand coastline, this wasn't to be. We ran out of both time and money, but we did make some orca observations from a plane and a helicopter. We were in Kaikoura and hoping to film orca hunting dolphins. This dramatic behaviour hadn't been well documented on film before and I knew it was happening off Kaikoura, especially in the summer months when the dusky dolphins were calving.

One day we heard on the radio that there were orca just out from the boat ramp. We raced off and found them not far from where they were first reported, and followed them for nearly six hours watching them taking rays. It was Betsy, Prop, Ragged Top, Bent Tip and Danny (the orca named after my research assistant), but they didn't hunt any dolphins while we were with them. It was a beautiful summer's day and as they headed off south of Kaikoura, Andrew suggested that we look for them further to the south the next day. Given that the coastline below Kaikoura is straight and without harbours, if the orca were sticking by the coast they should be reasonably easy to find. So off we headed in a plane at first light the next morning.

It didn't take us long to find them, and they were still tracking steadily south. It was incredible to see them surfacing all at the same time. From the air you could see nearly the whole animal at one time, and the size difference between the adult males and females was pronounced, with Prop being markedly smaller than the others. It was the first time I'd seen orca from above and it gave me a whole new perspective on their behaviour, as from the air you could get glimpses of them underwater when they wouldn't have been visible from a boat. However, as we watched them it became apparent that they weren't hunting so we left them and turned back to Kaikoura.

A few days later we had a report of orca off the Bluffs, to the south of Kaikoura, and this time they were heading north. I took off in my boat and Andrew and the film crew set off in a helicopter. Just before I arrived, the film crew saw the orca chasing a single dusky dolphin, but it escaped and headed back to the rest of the pod. It may have alerted them to predators being in the area, because they stayed a long way from the orca for the rest of the day. It turned out to be the same five orca as a few days earlier and I watched them hunting for a shark instead. Unfortunately, the film crew missed this hunt as they had already flown back to land. Overall we just didn't have any luck while in Kaikoura, and it turned out to be a season with one of the lowest orca-sighting rates for years. It was a shame, but that's the way filming wildlife goes.

One other event that occurred while filming has had a lasting effect on me. I had received a call that there was a lone orca seen off the south-western coast of the South Island. Two commercial fishers had seen it a number of times and brought it to my attention. Curly and Vanessa James, owners of the fishing boat *Nemesis*, assured me they knew they were looking at a very young orca calf, all on its own. I told them I was on my way.

I was over 1500 kilometres away, at my Tutukaka base, but I was going to try to get there before something happened to the calf. Danny and I started driving within an hour of hearing about it and kept driving for two whole days. We had my boat behind us, as we would need it to get out to where Curly and Vanessa were working. Arriving at Milford Sound, we met up with the film crew and immediately set out for the area where the orca was last seen. As we arrived, we came across Vanessa and Curly on *Nemesis* and they informed us that the calf was still there and took us over to the spot, near one of their crayfish pots.

Swimming around the buoy and rope we discovered a tiny little calf, about two metres long and probably weighing less than 180 kilograms. It was incredibly emaciated and its skin was all wrinkled from dehydration. One thing I noticed immediately was an algae growth covering its body, and that the skin on both the lower and upper jaws, near the rostrum, was red and damaged. But the most horrific thing, after seeing how skinny it was, was the three wounds we noticed – two punctures and one gash – on its right side. On closer inspection I could see that these were bullet wounds. Someone had tried to kill this calf and had injured it substantially. I was sickened to think that anybody could do such a thing, and to a baby no less.

I had tears in my eyes as I swam with this little orphan, which I could see from the pigmentation under its belly was a female. She stuck right beside me and swam where I swam, as if realising that her mother wasn't around and I was the next closest thing. Whenever I went back to the boat for some equipment or to talk to the film crew she returned to the cray-pot and hung around there. When I returned to the water she came straight back over to me. She was in dire need of liquid and food, so I tried to feed her some freshly defrosted fish, courtesy of Curly and Vanessa. She mouthed the fish, but did not swallow, then dropped it and picked it up a number of times, but after approximately five minutes left the fish to spiral to the bottom. When I returned to the boat the calf followed, but once we were under way she went back to circling the cray-pot rope.

We had bad weather the next day, and because my boat was so small I couldn't reach the area where the calf was. I decided to call her Nemesis, after Curly and Vanessa's boat – it seemed an appropriate name, given that she had been shot by someone. I found her the following morning, several kilometres to the south of her previous location. Again she came right over to me as I soon as I entered the water, and she was very curious about her reflection in the camera housing. I decided then that I would try and move Nemesis to a location where I could look after her better and try feeding her.

As we left that day I tried to formulate some plans in my mind. The best option would be to move Nemesis well to the north, to the sheltered bays of the Marlborough Sounds. I knew orca came into that area, so perhaps there was a chance any remaining members of her group might call in there; but if not, I also knew there would be enough rays for her to survive on if I could get her to the stage where she could capture them herself. I would probably have to helicopter her out, and catching her in the first place might be a bit of a mission, but I was sure it could be done.

Unfortunately, the next day there was a storm and I couldn't get out to sea. This lasted for four days and when I went back to search for her I couldn't find her again, nor on the following day. I never did find her and that failure continues to haunt me to this day. Might she have survived if I'd attempted to move her sooner? If I had spent more time in the water with her and tried to feed her more often, might she have taken enough fish to make it through the storm? I'll never know, and she will always live up to her name and continue to be *my* nemesis, by reminding me that I could have tried harder to save her life.

But near the end of our filming days I did get the chance to save another orca life, and that was Ben's. When we were first planning the documentary I'd told Andrew we'd get an orca stranding some time during the eighteen months of filming. Andrew may have been a little sceptical of my prediction, but as it turned out we did get one and, as I described earlier, we had the good fortune to save Ben near Whangarei Harbour. Then the very next day he was filmed back with his group. It was Andrew who filmed him, and unfortunately, for the first time in the whole period we worked together, I wasn't there, as I was just too exhausted and had stayed home to sleep. In the final cut they rewrote the ending to give the impression that I had been out there that day. It was a shame they did this, as it didn't really happen that way, but I suppose you could say that I had identified Ben from the footage so I was very much there in spirit.

# CHAPTER TEN
## The Battle of the Thesis

Throughout the whole time I was working with the film crew I was also trying to get my PhD thesis written up, and that included all the data-entry and analysis. My research assistant Danny Mayson was a hard worker and spent long hours making notes from my dictaphone recordings as well as helping out with the film crew when they needed a stand-in. It was fortunate that she's blonde-haired like me, because she could drive my truck, with the boat behind it, up and down the Kaikoura coastline as the crew filmed her (pretending to be me), while I was back at base working on the write-up. I had never found writing easy and now it was even more difficult. I had to be precise and accurate. There could be no wishy-washy aspects. This was hard science, and if I didn't do it right I might not pass. Although I knew that no one else knew as much as myself about the New Zealand orca, I still had to be able to articulate that knowledge effectively on paper.

As I was the only researcher working on orca in the South Pacific (and this still remains the case), I was going to have to compare my work with that done in the Pacific Northwest, Norway, Argentina, and anywhere else I could obtain results from. The examiners would be taking a hard look at how well I'd thought out my research plan, the results I'd obtained, the conclusions which I'd drawn from those results, and how my work compared with research carried out elsewhere. This wasn't the sort of thing you could pull together in an afternoon. I slaved on it for months and months, spending a lot of time in the library hunting down relevant scientific articles, and firing off countless emails to other orca researchers around the world, corresponding with them about their findings and results.

At the end of the day, I had a huge tome of a manuscript to submit. But the process was far from over, as now I had to wait anxiously for the examining committee to read it, make their analysis, and then inform me if I'd passed or would have to do a rewrite. There was also the oral exam, at which I would be challenged about the research project, or some other aspect of the thesis, and I would be given the opportunity to defend my findings. Typically, these defences take several hours, and the committee is comprised your supervisor, an external examiner, and an overseas examiner.

Fortunately, my supervisor John Craig had been out with me on the water and had seen how I conducted my research, and was in full support of it. My overseas examiner John Ford, the orca acoustics expert from Vancouver, couldn't attend in person but had sent in his questions. Both these examiners had valid and fair questions to put to me. For instance, John Ford queried why I was calling these animals orca instead of killer whales. I explained that this was how they are commonly known in New Zealand, and when it was suggested that I somehow substantiate this, it was agreed that a survey would be appropriate. So that was at least one more thing I would have to do before I could obtain my doctorate.

Then the external examiner (who shall remain nameless) started with questions. I was handed a sheet labelled 'Major Issues' which contained an itemised list of over fifty objections. My mind went blank. *Fifty* 'Major Issues' that were wrong with my thesis? Working through the list, I could see that some of the challenges were definitely valid – for instance, why hadn't I done more acoustical work with the orca in my study? This was an objection which I could address right there and then.

Although part of my original proposal had been to include an orca acoustics section, I explained that over the six years of field research for my PhD I had simply not amassed enough recordings to make a valid data analysis (an assessment with which John Ford, the world leader in orca acoustics, had concurred). There had been many times when I had tried to make recordings, yet the New Zealand orca just weren't making any noise. Only the week before, I'd been out with orca and after having spent more than five hours with them I hadn't managed to get a single recording of any calls. It wasn't a case of not trying, but that they just weren't talking.

Apparently this response wasn't good enough for the external examiner, who went on to object that if I was proposing that the New Zealand orca were so different from other populations, then I needed to do the acoustical work to show whether they had dialects such as John Ford had found. There was really no way to respond to that: I just hadn't collected enough data for this type of analysis. Although a typical PhD fieldwork time-frame was four years and I had already done six, this examiner was now proposing that I should have done more. I was totally knocked back.

But it didn't end there. The external examiner went on to say that any photos I had used from the public were not valid for inclusion in the analysis either, and that I would have to do all the photo-ID analysis again, including rewriting that chapter in my thesis to incorporate all the grading systems I'd used. I considered the addition of the grading system a reasonable request, but I certainly didn't agree to not being able to use data collected from the public.

This method had been used for the very first orca study (by Mike Bigg) and was an accepted method of data collection for that project, which was continuing to this day. In addition, I had detailed records of who all the photos were from and their degree of reliability. Who was this person to decide that the public wasn't good enough to be involved with orca research, and that my thesis – to quote the examiner's exact words – was 'of no scientific value'?

After nearly seven hours of grilling, I walked out of that exam room totally deflated. I was distressed, and even more so, bewildered. If the other two examiners had found so little to comment on about my thesis, then where did this litany of objections come from? Could they be justified? Or was there more to this than met the eye? I spun around right there and then, and marched back to John Craig's office to confront him. He just smiled and said, 'Now, that wasn't so bad was it?' I had to laugh. Everything was going to be all right, somehow.

So John and I started working on a game plan. We agreed that there certainly were parts of the thesis that could do with a rewrite, so if I was going to do this, then why not prove the external examiner wrong at the same time and publish them scientifically? The peer-review process involved was going to be lengthy but it would clearly show my findings were valid, and once published I could quote them in my thesis and the external examiner would have no grounds to object to the 'scientific value'. This was a brilliant idea, so off I headed home to start writing immediately.

The next six months were taken up with addressing some of those issues brought up in the exam. First I conducted surveys to find out whether 'orca' or 'killer whale' was the most commonly used name in New Zealand. I didn't want the public to just tick a box, I wanted them to give me the name without prompting. Therefore, I devised a sheet made up of diagrams of marine mammals commonly found in New Zealand waters; people filling out the survey would have to write the animal's name beside each drawing. The pictures of a New

Zealand sealion, a humpback and southern right whale, a bottlenose and common dolphin, a Hector's dolphin, and of course an orca weren't in any particular order but were scaled to size. Surveying three different age groups – primary-school kids, high-school students, and adults – I obtained results which supported my assertion that orca were well known by people generally, and that the majority of them called them by that name. The adults actually had the lowest scores, but even then ninety-four per cent of them could name the picture of an orca as either an orca or a killer whale, and it was ninety-eight per cent for high-school students and an astonishing 100 per cent for primary-school kids! Across the age groups more than fifty per cent of them used the name orca – one even named the orca as Free Willy. The findings were conclusive and I added a new section to my introductory chapter.

Then I started taking other sections of my thesis and rewriting them. Where appropriate, I turned the sections into manuscripts which I submitted to various scientific journals. In a little over ten months I had seven manuscripts accepted for publication, covering many of the aspects of the lives of New Zealand orca which I write about in this book. I had also rewritten nearly all my original chapters and included completely new sections on the associations of the orca – such as who was hanging out with whom – and relating those findings to who was food-sharing with whom and in what context that behaviour had occured. My aim was to show how these 'Association Indices' – i.e. the measurements of how much time orca were with, or without, each other – were relevant to other aspects of the lives of New Zealand orca; specifically, that there was persuasive evidence for the New Zealand population of orca having three distinct sub-populations. Arguing that Association Indices could be used to look at population structure, and giving evidence of their use by whale and dolphin researchers in different contexts, I set about applying them to identify the sub-populations of New Zealand orca. Below is a small sample of the kind of complicated analysis I was required to submit on this topic. (Just skip over this stuff if you find it too boring or heavy going!)

Association Indices are a measure used by researchers to support the existence of potential group structures within a population. They have been used by cetologists for a variety of species, e.g., humpback whales (*Megaptera novaeangliae*) (Clapham 1993), bottlenose dolphins (*Tursiops truncatus*) (Ballance 1990, Bräger et al. 1994, Conner et al. 1992, Schneider 1999, Smolker et al. 1992), Hector's dolphins (*Cephalorhynchus hectori*) (Bejder and Dawson 1997, Slooten 1990), spinner dolphins (*Stenella longirostris*) (Östman 1994) and orca (Heimlich-Boran 1986, Heimlich-Boran 1988). Results from PNW orca show that some individuals are regularly sighted together, whereas others are never sighted together, and may avoid each other (Baird 1994, Ford and Ellis 1999, Ford et al. 1994).

In social animals kin are often seen together (Fletcher and Mitchener 1995), e.g., chimpanzees (Goodall 1991), gorillas (Fossey 1974, Fossey 1983), baboons (Cheney 1978), hyenas (Holekamp and Smale 1990), swamp hens (Craig and Jamieson 1988), elephants (Moss 1988) and orca (Ford et al. 1994, Heimlich-Boran 1986). These associates often participate in joint activities, including playing (Cheney 1978, Fossey 1983, Goodall 1991), assisting in rearing young (Craig and Jamieson 1988), alloparenting and 'baby-sitting' (Bisther and Vongraven 1993, Bisther and Vongraven 1995, Vongraven 1993, T. Similä pers. comm., Visser, unpubl. data), care-giving behaviour (Ford et al. 1994), carrying deceased (Conner and Smolker 1990, Schneider 1999) and food sharing. Food sharing has been reported for various taxa, e.g., hyenas (Holekamp and Smale 1990), lions (Schaller 1972), chimpanzees (Goodall 1991), gorillas (Fossey 1974, Fossey 1983), swamp hens (Craig and Jamieson 1988) and for cetaceans (whales, dolphins and porpoises), e.g., delphinids (Fertl et al. 1995) and orca (Baird 1994, Baird and Dill 1995).

New Zealand orca appear to have a small population size and the data suggests that certain animals use different parts of the coast and hence may form distinct sub-populations. Some individuals are regularly sighted together, display alloparenting and baby-sitting behaviour and have been observed sharing food. This chapter seeks to elucidate population structure (age and sex ratios) and, through the use of Association Indices and food sharing between known individuals, evaluate the likelihood of these population sub-structures.

As you can see, this was pretty heavy-going stuff to read, and believe me, it was even more taxing to write. And the results were equally as complicated. I had to present my findings – which of course I knew intimately because I had been working with the animals and collecting the data over many years – in a way that any researcher, even somebody who might never have seen an orca, could understand. It was an enormous challenge. I ended up with a huge Association Indices table (with 117 animals across the top and 117 down the side) that was just too big to deal with. So, deciding that I would only look at the animals I had photographed five times or more (which would give me a good sample size and also eliminate animals from the data analysis that I didn't have much information about), I whittled the field down to fifty orca who had been seen more than five times and who had also been seen with at least one other orca.

But now I faced the added challenge of trying to set forth those results in a way that would represent the group structures, if indeed there were any. Given my difficulty with numbers, I wanted to find an easier way to show if the animals mixed, and how often they did it, than just the numbers from the Association Indices. Using a similar system to that proposed in an Hawaiian PhD thesis to describe the social organisation of spinner dolphins, I created a 'circle-plot' with the codes for the orca (e.g. NZ1, NZ50) placed around the outside and lines of different thicknesses drawn between them to represent the Association Index value. In other words, the higher the Association Index (how much time each orca spent with another animal) the thicker the line, and vice versa.

I also wanted to show if the orca had been photographed only off the North Island, only off the South Island, or off both, so I drew a line which was placed either under or above each orca's code number. The line represented Cook Strait, the body of water separating the two main islands, and if their code was above the line then the orca were seen only to the north of Cook Strait – i.e. they were North Island orca – and likewise if their code was under the line, they belonged to the South Island. If they had been seen off both islands they didn't get a line at all. Initially it took a fair bit of shuffling around, but eventually everything worked out quite well. (Further discussion of my Association Indices methods and circle plots are given in Appendix Two, pp. 231 and 232.)

With all the groundwork now firmly in place I could set about writing up the results and interpreting them in the context of what I had actually observed when out with the orca – for instance, if two orca, say A1 and Olav, had an Association Index of 0.93 this meant that they were seen together nearly all the time. Overall, I found that there were some very distinct groupings of orca – those who spent considerably more time together than others, and those who mixed infrequently. Additionally, I could see that some orca (such as Ben), who had only been seen off the North Island, were seen with orca who had been seen off both the North and South Islands, so there did appear to be some mixing going on – or else there were large gaps in my home-range data on some of the orca.

Now that I had compiled all these results (along with those for the food-sharing and the sub-population hypothesis), I had to put everything into the context of a theory. Why exactly was this stuff happening? What could it mean, and what follow-up steps should be taken to expand the research?

At the same time, I outlined that there were less than 200 orca to be found around New Zealand. The early trends I had discovered in my pilot project remained the same even after six years of added data. There were clear distinctions between the areas that the individual orca were sighted in. The results indicated that there were definitely three distinct sub-populations – one that was found only around the north of the North Island, one that was found only around the South Island, and another that travelled all around the country.

This would mean that the population estimate I had come up with, of less than 200 orca around the *whole* of New Zealand, should be considered as smaller units, as it was divided into these sub-populations. If they *weren't* mixing then there could be all sorts of potential problems such as not enough breeding females or a high risk of extinction in the event of an environmental disaster such as an oil spill. Alternatively, if these different sub-populations *were* mixing then the population would be larger so might be more stable.

I added to the weight of this theory by using the few skin samples I had collected from orca, which indicated the New Zealand orca were most likely divided into at least two different sub-populations. Due to the limited number of skin samples, I couldn't be sure about the third sub-population, but even dividing the known number of orca into two different groups (and who was to say that there was an even split between the two?) gave a very small genetic group, no matter how the population was distributed.

For the discussion section of this part of my thesis, I wanted to be able to show the connection between the amount of time orca spent with one another, the food-sharing, and the links all this had to the postulated sub-populations. If I could show that not all the New Zealand orca mixed, then we would have to treat each sub-population as a separate entity and we could start looking at ways to protect them. The different sub-populations could well be subject to different threats, and to lump them all into one homogenous group would not be an effective method of protecting either them or their habitat. For example, orca living off the North Island spent a lot of time inside harbours and estuaries hunting for rays, yet these types of habitat weren't protected. Orca seen off the South Island didn't have large shallow harbours to hunt in, but they might be heading down to Antarctica to hunt for penguins and so would need protection from whalers.

Pulling everything together with a chapter on the conservation management of New Zealand orca – what things were threatening them, or were likely to in the near future, and what we could do to help eliminate these threats, which ranged from the possibility of them being hunted by whalers, to oil spills, to collisions with boats – I submitted a strong argument for putting in place appropriate measures for the protection of our unique orca. (Considering that one of the main North Island hunting grounds for rays was Whangarei Harbour, which contains an oil refinery, an action plan for preventing orca entering an area where an oil spill had occurred was a particularly important aspect to take into account.)

Eventually I had a completely new version of my thesis and I was pleased with the results. I may not have agreed with many of the reasons behind the external examiner's thinking, but the outcome was definitely worthwhile. All I could do now was sit back and wait for another couple of months while the examiners read the revision. I was confident that this time things would go more smoothly and that it would be a relatively painless exercise.

Silly me, I should have learned by now that nothing useful in life is easy. Again, I received good feedback from both Johns, but the external examiner was spoiling for a fight. There was a second list of 'Major Issues' – nearly as long as the first one. John Craig was not impressed and he arranged for a mediator to be present. This was going to be my last chance to defend the thesis, and if the examiners couldn't agree on the results I wouldn't be able to submit it again. I could try and work with another university, but I didn't have the money to start all over again. I was totally floored and remember just sitting in the room without reacting. I couldn't believe the system could allow this sort of thing to happen.

In the end, after several more hours of wrangling by John on my behalf, the mediator stepped in and instructed the external examiner to sign the papers to approve the thesis. Finally it was signed and I didn't even feel relief; I was so numb with shock that I just walked out to my car like a zombie. I don't even remember the drive home, or the next couple of days. I was issued my PhD certificate with only a few further hassles. It had been a mission getting there, but I had finally done it.

Two years after my thesis was finally published, the Department of Conservation issued a new classification listing of all the native species of plants and animals. Based on my recommendations, they had altered the classification of New Zealand orca from 'Common' (a classification only ever plucked out of thin air) to the highest category of 'Nationally Critical'. Amending their status might impose some moral pressure on the government to act accordingly now that orca were to be listed in New Zealand's highest possible threatened category, which was equal in status to 'Critically Endangered' in the IUCN Red Data listing.

However, DOC didn't get it completely right, as they had some qualifiers. These were that orca were 'Secure Overseas', which we know not to be the case (some orca populations were in decline and others hadn't even been studied so couldn't be designated as 'Secure'), and that the New Zealand population was 'Stable' (which again we don't actually know). So the orca were finally classified into an appropriate category but with erroneous data. It looked as if I was going to have to keep trying to change this. For one thing, now that orca had gained a threatened status within the local system it was important to get this recognised overseas. Getting the New Zealand orca registered officially in the Red Data List would be one of my next missions.

# CHAPTER ELEVEN
## Adopt an Orca

I wanted to keep working with orca long term and I needed to find a way to earn money to keep the research project going. Although I had been working on eco-tourism ships and doing a few other odd jobs, it was impossible to hold down a 'normal' job and still have the flexibility to up and leave whenever I heard there were orca in the area. A job which involved orca, if at all possible, would be ideal. So I thought I would start an 'adoption' programme – one in which people contribute to the research by supporting it financially, as well as getting something in return such as newsletters or certificates. I hoped it would also be a good means of getting important information about orca out to the general public.

Similar programmes had been set up overseas, so I contacted a few of them and enquired about logistics. I wanted to be sure that something like this could be run by myself and that it would be worthwhile. I also wanted to find out how they ran the actual adoption side of things. Depending on where the project was run from, and the type of cetaceans they were offering up for adoption, there was a wide variety of formats. None of the programmes supplied a map of where the animal had been sighted, so I decided that this was something I would like to do. In fact, there were many things I picked up from these other adoption programmes, and perhaps the most important was that there was only a small market out there, even within a large population base. The other was that, in reality, such a scheme involved too much work for just one person.

So I got a couple of other people on board. One of these was a woman involved in diving who was also very keen on orca, but had never seen one. The other was a retired housewife I'd met at Ben's stranding and rescue, who was subsequently keen to help orca. I thought I now had the perfect team – someone who was into the marine environment and someone who was retired with all the time in the world – so we started having meetings and getting 'Adopt an Orca' under way.

Showing them my business plans, I explained how the system would work, including things like the sightings maps and educational packages. I paid to have trust documents drawn up and the programme registered as a non-profit trust, and had a logo designed. We looked at various items we might be able to sell through the scheme, who our target audiences were, and how the newsletters would work, and I suggested that one of them could run the membership and the other the educational side of things. Meanwhile, I would undertake the research and supply all the necessary information and photos for newsletters and the website. We got a website address registered and weren't far off launching in time for the 1998 Christmas market.

Then I heard that I had been sponsored to go to a marine mammal conference in Europe to present my orca findings. This was an incredible opportunity to not only share my findings about the New Zealand orca, but meet other orca researchers from around the world.

This is Ben when he stranded. He was talkative that night and was whistling and chirping away. He is lying on a special mat, which we later used to attach to a rescue pontoon system. **Terry Hardie**

After being on the beach all night Ben was keen to get back out to sea and, once we had him in shallow water, he helped us by using his strong tail. He was then placed between rescue pontoons – almost like an inflatable boat without a hard bottom. Once out in deep water he was released and was sighted with another orca only an hour later. **Terry Hardie**

One day, just off the coast near Tutukaka, I had a fantastic encounter with a group of orca. Here, a sub-adult male and I were in the water eye to eye. Orca have the same number of bones in their flippers as we do in our hands, but they can't move them because they are all embedded in the rigid flipper. Perhaps, through their echolocation, they can determine that we have similar bones, as he seemed fascinated with my dexterity as he watched me click my fingers. **Jen Schorr**

This is one of my favourite underwater orca photos. The sub-adult male could see himself in the reflection on the dome of the underwater camera housing. He lay there watching himself, but when I went back to the boat for a new film, he followed me and waited beside it until I returned to the water. **Ingrid Visser**

Adult male orca have pectoral fins which are just huge – these perfect examples belong to Keiko the star of the orca movie Free Willy. Females have much smaller pectoral fins, which are about a third of the size. **Ingrid Visser**

Keiko was an amazing orca who touched the hearts of all who met him. I was fortunate to get the chance to work with him in Iceland, when he was being trained for release back into the wild. I have a training whistle in my mouth and am on a platform hanging out over the water of his pen – a fenced-off bay.
**Terry Hardie**

Here an orca and I blow bubbles at each other. This young female was very interested in what was going on in the boat and kept coming over for a look. She would blow bubbles each time she visited, so I decided to do the same thing. Next she rolled over and hung upside down, right next to me.
**Jen Schorr**

Orca are the top predators in the ocean. They are sometimes called apex predators – meaning they are at the top of the food chain. They accumulate poisons and toxins from all levels in the food chain, and have limited ways to offload them so can become very sick. Therefore they are also indicator species – they can indicate if there is something wrong with the marine environment lower down in the food chain. This is Rocky with Venus in front of the oil refinery at Whangarei. **Ingrid Visser**

Even dead orca can reveal interesting data. Here I am kneeling next to a skull from an Argentinean sub-adult male. If this skull was studied we could tell how old he was (by counting the rings from a section of a tooth), and also ascertain his genetic links to other orca from the area and how closely related his population was to other orca populations around the globe. **Gretchen Freund**

The only underwater photograph I know of which shows an orca and a mako shark together. Miracle, the orca closest to the shark, had it in her mouth when I first got into the water. She let the shark go after a while, and it came and hid under me. Miracle chased it out from under me and then proceeded, with the other orca just peeking into the picture on the right, to eat it for lunch. This was the first record of orca, anywhere in the world, eating a mako shark. **Ingrid Visser**

This group of orca is swimming along a beach well known for rays – prime hunting ground, but also a prime stranding area. More than five strandings have been recorded on this beach. The adult male in the front of the group, Rua, stranded in 2001 on a different beach. Luckily for him the local rugby team was nearby and helped to re-float him. It was three years before we had a re-sighting of him and knew that he was okay. **Ingrid Visser**

The perfect sunset. This was taken in the Bay of Islands without a filter. The orca on the left is Spike, an adult male, and the tail belongs to Nobby, a sub-adult male. **Ingrid Visser**

Peninsula Valdez, in Argentina, is famous for its orca who come up onto the beach to take sealion pups. Here two females are patrolling the beach, while the wary adult sealions watch. **Ingrid Visser**

Most people don't think of orca living in tropical waters, but this photo shows a female with palm trees in the background. It was taken in Papua New Guinea waters. I have been there five times looking for orca. During one encounter I watched them eat a hammerhead shark. **Ingrid Visser**

Here I am with Rudie who stranded himself at Taiharuru, Northland, near Whangarei. You can see just how big orca are in this photo – Rudie is about 6.6 metres long. **Mike Cunningham**, *Northern Advocate*

Calves closely accompany their mothers on most surfacings and dives. When travelling the calf rides in the slipstream of the mother, increasing the drag for her, but making it easier for the calf to keep up, as well as reducing the amount of energy it needs to swim long distances. On some deeper dives the calf may stay nearer the surface, but it is rarely alone, and often has a babysitter. **Ingrid Visser**

I carried on working on the Adopt an Orca project, but as I was preparing a presentation for the conference, as well as putting the finishing touches to my thesis at the time, something had to give. The best thing to do would be to put Adopt an Orca on hold for a couple of months until I got back from Europe. Based on projections from results of similar programmes overseas, I tried to convince my partners that if we waited to launch for Easter we would still be okay.

However, showing them the projected financial outcomes turned out to have been a very silly thing for me to do. Two days after I arrived in Europe I got an email to say they were going to set up their own adoption programme. They seemed to think that they could launch a similar programme and earn a full-time wage from it for themselves while just offering grants to researchers annually. Their reasoning was if that they didn't have the overheads of actually conducting research to worry about they could run an adoption scheme at a lot less than my projected budget. Worst of all, they had a copy of my orca sighting database so they had all the information they needed to set up.

I was devastated – did these people have no integrity? I had spent years carefully researching and planning how this was going to work, and now they had taken all my ideas, data, the logo, the business plans and all my orca information and were going into competition with me. It's not that I was afraid of the competition, as I knew that nobody in New Zealand could offer any updated information on the orca, so that aspect of their programme was bound to collapse. What I was concerned about was that the New Zealand market was only big enough for one adoption programme – my projections for Adopt an Orca had clearly shown that. Additionally, I knew that I wouldn't have time to get my own programme up and running before theirs if I didn't go back to New Zealand and sort it all out. Not knowing what else to do, I phoned my dad and spoke to him at length. He suggested that I continue on with the conference and he would have my lawyers send them a letter.

# ADOPT AN ORCA

I arrived home shortly afterwards to find that the letter had been sent and that I would get all my orca data back. They would have to change their logo and the trust documents. They couldn't use the word orca in their name or put up any orca for adoption. No matter how much work was involved, I was now determined to get my own programme up and running. So off I went.

I had new logos designed, trust documents written up, and trustees appointed. I had adoption brochures and certificates printed. I conferred with Bob Owen of Design Online, who offered to sponsor the design of the website, and my computer guru friend Terry Hardie offered to host it. I wrote up the personal histories of the orca who were going to be up for adoption and I laid out maps showing where they had been sighted. I spent hours with a computer database designer, Paul Prosée, who was sponsoring the database. I designed and had printed postcards with orca on them which would be given away as gifts when people signed up. I sorted through hundreds of photographs for the right images for the first newsletter, which I also designed, wrote and then had printed. I was in constant contact with the Whale and Dolphin Conservation Society (WDCS), based in the UK, who ran one of the largest adoption programmes in the world and had helped financially supported my research during the last years of my doctorate.

Then one day in the midst of all this frantic activity I cleared my mail to find a letter from Project Jonah informing me that they were now working with a newly formed organisation which was going to offer whale and dolphin adoptions. There was going to be a launch in a month and they wondered if I'd be interested in giving a talk about my orca research at the launching party. At first I couldn't believe my eyes. What a cheek. And then I realised that the Project Jonah committee members probably had no idea of what had happened and had contacted me in good faith. And if I looked at it another way, here was a golden opportunity to show what I had done, not just with my research but with Adopt an Orca as well. I wrote back to say I would be delighted to give a talk about my research but that I would also require a display table for some of the items I had on offer to support ongoing orca research. This was all confirmed and four weeks later I was to speak at the launch of the rival whale and dolphin adoption programme.

In that intervening month I launched my Adopt an Orca programme. I contacted the media, articles appeared in magazines and in newspapers with by-lines such as 'Have a pet that won't pee on your couch', and cartoons showing kids bringing a large orca home to put in their swimming pool were syndicated nationwide. I was invited to participate in radio talkback shows and even appeared on a few TV current affairs programmes including the top-rating *Holmes*. I was receiving good feedback and processing adoptions as quickly as they came in.

Now it came time to go to the launch of the competition's scheme. When I arrived I was disappointed to find that they were actually charging the public a fee to come in and see their launch. If they were asking people to donate money to support whales and dolphins, surely they wouldn't charge them to come in the door *and* make a donation? But the door fee wasn't the only thing that astonished me; they were totally unorganised in myriad respects. I wasn't the only researcher who'd been asked to give a presentation, but the audio-visual equipment wasn't working properly. Eventually I headed out to my vehicle and brought in my own slide projector sponsored by Kodak and things went smoothly from there on in. When it came time to give my presentation I talked about the methods I used in my orca research and some of the results I'd obtained. I finished up by telling the audience that I had an Adopt an Orca programme which anyone who was interested in could find out more about at the display table where I had orca information available for them to take away.

My sister, Tanya Jones and Sarah Kessell, an accountant and trustee for Adopt an Orca, had come along to help out. We were all wearing matching Adopt an Orca T-shirts and caps and the display table was covered in Adopt an Orca goodies. Adoption brochures were all ready for people to sign, and sample adoption packs were available to view. We had miniature orca hanging from a fake Christmas tree to show that adoptions could be given away as gifts, and large blow-up orca hanging from the ceiling above us as well as Adopt an Orca postcards to hand out. We had it all and it all looked great. Dare I say it, it looked even better when compared with the 'new' whale and dolphin adoption programme display table which had a sign up saying that their merchandise was not yet available and their brochures were at the printers.

Eventually the rival adoption programme did get off the ground and after a few years it even started giving out money to researchers, although I have yet to receive any funding from them for orca research. Further down the track they also put up an orca for adoption – Ben – as he was the only orca they might have a chance of identifying from information supplied by other whale and dolphin researchers around the country. Interestingly, they recently approved a grant to support shark research – which is remarkable given that they are a whale and dolphin adoption programme.

Adopt an Orca continued to run well for a number of years and certainly helped raise the profile of the New Zealand orca. Nearly half the adopters were from overseas and lots of people adopted more than one orca. The newsletters proved very popular and were a great way of getting information out to 'orca parents' as well as people who had given adoptions as gifts. Information from the newsletters also appeared in other orca adoption publications, such as those put out by the Wild Killer Whale Adoption programme based in Vancouver and the WDCS Whale Adoption programme in the UK and Australia.

As well as supplying fuel for the boat and four-wheel-drive vehicle when doing orca research and other ongoing costs like film development, Adopt an Orca paid for various one-off items such as a cover to protect the boat from the sun and educational aspects such as funding the publication of various scientific articles. It was also during this time that I met, aboard one of the eco-tourism ships I worked on, Tom and Greta Newhof from the United States. They offered to help me set up a sister organisation, Adopt an Orca USA, which would allow US citizens to take advantage of their tax laws when making donations. Through their assistance we received some healthy contributions.

But overall there was a lot of work involved and Adopt an Orca ultimately reached the stage where I couldn't run it by myself. I had Chrissy, my step-mum, doing all the memberships for a while, but there was an enormous amount of work involved in sending out information to prospective adopters, writing the newsletters, getting memberships entered and renewals sent out, not to mention trying to conduct my research as well. Therefore, after running for a good six years, Adopt an Orca eventually reached a point where administration was taken over by the Australian WDCS branch, although the US sister branch is still run by Tom and Greta. WDCS Australia already supported an adoption programme for dolphins so they were well versed in running such projects. For me, it meant less time spent on paperwork, and therefore more time to spend getting out in the field and studying the orca. Even better, with their help Adopt an Orca grew in leaps and bounds and hopefully will eventually raise enough money to support other orca research programmes.

# CHAPTER TWELVE
## Where to Now?

While writing up my PhD, the scientific papers, and setting up Adopt an Orca, I had been spending more and more time behind a computer and less and less time out with the animals. I wanted to find a way to remedy this, but I still needed to earn a living as well. Although I could get some financial assistance from funding applications and Adopt an Orca, that wouldn't take care of my ordinary everyday living costs.

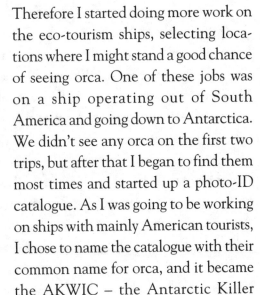

Therefore I started doing more work on the eco-tourism ships, selecting locations where I might stand a good chance of seeing orca. One of these jobs was on a ship operating out of South America and going down to Antarctica. We didn't see any orca on the first two trips, but after that I began to find them most times and started up a photo-ID catalogue. As I was going to be working on ships with mainly American tourists, I chose to name the catalogue with their common name for orca, and it became the AKWIC – the Antarctic Killer

Whale Identification Catalogue. In the same way I'd done for my New Zealand project, I asked people to send in their sightings and photographs, and it didn't take long for me to have nearly 100 orca in the catalogue. This may not sound like many, but when you consider that most people visit Antarctica for a ten-day trip and never see them, it was a good start. In fact, I knew people who had been working on eco-tourism ships for over five years and had never seen an orca in Antarctic waters.

Then in 2001, working with Jo Berghan, I put together a presentation on the AKWIC for a marine mammal conference in Vancouver, and this generated further responses from the public and from scientists who had been down to Antarctica working on other species of cetaceans. By now I was down in Antarctic waters for an average of a month a year. Not a lot of time, but more than I could have managed any other way. I had seen all three types of orca (Type A, B and C) which could be found there and was starting to compile a reasonable database of sightings. Although most were from the Antarctic Peninsula region (where I was working), and predominantly during the summer, I was getting a few records from other areas and during other months of the year.

In November 2003, I was working on a ship touring the Antarctic Peninsula and we were observing a solar eclipse. In the middle of watching it, our captain, Philipp Dieckmann, spotted some orca. Manoeuvring the ship closer through the large swells, we had an incredible encounter with them as they rode down the waves right next us. I managed to get some good photo-ID shots, including one of a female with the top of her fin missing, just like A1. I could tell that it wasn't *her*, but they were certainly similar. We spent a magical hour with the orca as they bow-rode and surfed among the icebergs. And because we had encountered them during the solar eclipse, I called the female with the top of her fin missing Eclipse.

Two months later I was working on the sister ship to the one I'd been on in November. Late one evening there was a call from the bridge – they had seen orca. The expedition leader, Susan Adie, who knew about my research, had the captain turn the ship around and we went back for another look. We slowed down the ship as we approached, and the orca came right over and started bow-riding. The seas were pretty calm, so they couldn't surf, but they were playing around happily in the vessel's wake. The light was very low, but I did manage to get a few photographs. Then just as we were turning to leave I spotted an orca way behind the ship with the top of its fin missing. I was beside myself – could it possibly be Eclipse? I managed to get one very far-off photo of her as she disappeared behind the stern, then rushed down to my computer to download the images from my digital camera.

Sure enough, the photo was a long way off, but it was definitely an orca with the top of its fin missing and not just a trick of the light. I had the photographs of the orca from my November Antarctica encounter with me, so I started hunting through them. The keen whale and dolphin watchers aboard the ship (a group from WDCS in the UK) were all leaning over my shoulder as we browsed through the images. Suddenly I got goosebumps and a hollow feeling in my stomach – we had a match. It *was* Eclipse. After almost ten years of trying to get a match for Antarctic orca I had finally got one, and in both instances it had been me who had taken the photograph. I slowly calmed down and kept looking through the files, and eventually managed to match three orca from both encounters, plus a note that one of the females was accompanied by a calf each time. These matches weren't mistakes and they weren't mis-matches. These orca were the very same ones I had watched bow-riding in November.

I was so excited that I couldn't sleep that night, so I ended up writing the first draft of a scientific manuscript about the matches. These were the first ever photo-ID matches for Antarctic orca and I wanted to be sure that the information got out there. The sightings were seventy-seven days apart and there was only sixty-five

Working in Antarctica is one of the ways I earn money to support my research. I guide ecotourists and show them the wildlife and spectacular scenery. This photo was taken at Neko Harbour. Working down there also gives me the opportunity to look for and conduct research on Antarctic orca. From that research I have established a catalogue of individual orca, similar to the ones I have set up for New Zealand and Papua New Guinea. **Allen Davies**

kilometres (thirty-five nautical miles) between the two locations. What could this mean? Well, as there were three orca matched between the two sightings, it could be deduced that the animals had most likely stayed together during that seventy-seven-day period. Perhaps this was the first indication of group structures for Antarctic orca (but I was wary of designating just what that group structure might be, based on so little information). The short distance between the two sightings might indicate a number of things. The orca could have stayed in the general area between the two encounters – perhaps indicating a core area of their home range – or they could have left and come back. Seventy-seven days would have given them plenty of time to travel to another location and return. Basically, the photo-ID matches were providing data which was fuelling more questions than it was answering, but it was a start.

These Antarctic findings also prompted me to rethink my sightings of the grey orca in New Zealand waters. I wanted to illustrate that these encounters were not a matter of my mixing up the data, or one of mistaken identities. I was *sure* they were Antarctic orca. Part of the problem was that when I had seen the group of grey orca off New Zealand in 1997, I had also noticed some scars on a few of them. These scars were oval-shaped and healed over, but they were exactly like those I had seen on pilot, humpback and Bryde's whales, *Balaenoptera* sp. – and the current theory was that they could be attributable to cookiecutter sharks, *Isistius* sp.

These little sharks, which only reach about fifty centimetres in length, are voracious predators, making up for their small size by attacking creatures hundreds of times larger. They are deep-water sharks that only come to the surface at night to feed. When they find suitable prey, they attach themselves with a vacuum grip from their lips and throat, bite out a semicircular plug of flesh, then start spinning around, cutting out the plug completely. They then drop off the side of the victim and swallow their bite of meat. Cookiecutter sharks are so avid when hunting for their prey that they will bite anything they come across. They have even been known to take plugs out of submarines, and nowadays the rubber-coated sonar domes of subs have an extra layer of fibreglass to protect them from cookiecutters.

Cookiecutter sharks inhabit warm waters. If I could find similar marks on the orca when they were in cold Antarctic waters, it would suggest that those orca must have left the region at some stage in their lives. Additionally, there are no known animals like cookiecutters down there which could inflict such scars. And I got really lucky with one photograph of an adult male orca in Hoovgard Bay, Antarctica. It clearly showed him surrounded by icebergs in an Antarctic setting, so there could be no mistaking where he'd been photographed; and sure enough, on his side he had a big oval-shaped scar just like a cookiecutter shark bite. My sightings of 'Antarctic' orca off New Zealand were beginning to look like not such an impossible thing after all.

Perhaps the Antarctic orca had been leaving these waters regularly, but they just hadn't been seen elsewhere; after all, it was a big ocean out there and who knew where these orca really went? Nevertheless, I still had to take into consideration that orca had been seen in ice down off Antarctica in the middle of winter, so obviously they weren't all leaving, even though some sure looked like they were. I was determined to get to the bottom of this.

Following up on the lead of the cookiecutter bite-marks, I also wanted to go looking for orca in warm waters, because that was where these little sharks were found. Very little research had been done on orca in any tropical waters, and there was none at all done on the orca in the South Pacific, apart from my New Zealand research. I had heard that there were orca occasionally seen off the Walindi dive resort in Papua New Guinea (PNG), and that those orca had been seen feeding on sharks. If, as I suspected, orca develop their method of feeding by being taught, then perhaps the PNG orca had learned how to catch sharks from the New Zealand ones – or the other way around. Perhaps when I couldn't find orca around New Zealand, they were going to PNG? I didn't know, but I thought I had better go take a look.

To begin with, I spent a lot of time writing letters and trying to track down information on orca sightings from all over PNG. This was a difficult process, as over there communications are unreliable at the best of times, and in many areas the people can neither read nor write. As for email, there were only one or two locations in the whole country where that was possible. I contacted the owners of Walindi, Max and Cecelie Benjamin, to ask them what they knew, and they invited me to come and stay to see if we could find orca ourselves. Next I wrote to Frank Bonoccorso, the curator of mammals at the PNG Museum in Port Moresby, who had a few sightings and some information he would share with me. Now I just had to find a way to get there. Amazingly, an opportunity came along to work on a ship that was passing through PNG. I set off with high hopes, but once aboard I was told by the captain and crew that they had never seen orca in PNG, so our chances were pretty slim. Well, the orca-luck was with me, as only a few days into the trip I heard a call on the hand-held radio that the ship's first mate, Martin Breen, had spotted orca. They were moving fast and by the time I got in position I only managed to get three photo-ID shots, but I had the beginnings of the PNG orca catalogue.

On the way home from working on that trip I had managed to route my flights via Australia, and from there it was just a three-hour flight to PNG. Taking an internal flight to Walindi, I met up with Tammy Peluso, who'd moved there from the US after seeing orca for the first time when Max took her out diving at Kimbe Bay. She had started a photo shop at Walindi and her logo was an orca. Together we were determined to find the PNG orca, and went out for a few days on the dive boats looking for them, but no luck. I was astonished at the number and variety of other cetaceans we were seeing, but no orca.

One day Max asked if Tammy and I would like to take a boat out and go and search exclusively for orca, rather than just hitch a ride with the dive boats. Would we what! So off we set, Max, Tammy, myself and our local skipper Joe. Just as we neared the end of the bay and were about to turn for home I shouted, 'Orca!' – and I know that Max thought I was joking – but it was for real. There were six of them, and even from this distance I could see that one had something in its mouth. 'Quick, Joe, head over towards that one,' I yelled, as I pointed at a female who was hunched over at the surface, 'she's feeding on something!'

Well, as we got closer we could see that she had a scalloped-hammerhead shark in her mouth. This was fantastic and my orca luck in PNG had held out again. The shark was upside down in the orca's mouth, held exactly the same way New Zealand orca hold sharks. I got a photo-ID of the female with a large notch out of her fin and her calf swimming right next to her, and we watched as they shared the shark and then disappeared. It was only a quick encounter, but it was a good one. This was the first time a scientist had seen orca in PNG feeding on any species of shark, and not only that, this was a new prey species for orca worldwide. No one had ever seen orca feeding on scalloped-hammerheads before.

I was on a roll, and together with Tammy, started working full steam on the catalogue. With the photos she already had, the ones I had taken, and some that I'd collected from various other people, we made some matches. Frank Bonoccorso and I published the first ever scientific paper on orca in PNG waters – which was also the first ever scientific paper about any species of cetacean in PNG. I was flabbergasted – no one had ever researched any of the whales or dolphins in this area, yet here was this plethora of marine mammal life right outside the door. I was determined to come back, and immediately set about writing proposals.

So far I have been back to PNG three times and I hope to make regular surveys. I have now seen thirteen different species of cetaceans in the waters off Walindi, eleven of which are not even officially found in PNG waters (at least according to the IUCN Red Data List). I hope to publish another paper about all the other species found there and, together with The Nature Conservancy, a US-based conservation organisation, work towards getting Kimbe Bay declared a marine park. This will take a lot of work, but it's definitely worth it.

To date, I haven't seen any New Zealand orca in PNG waters, but I've only been looking for a few years. For that matter, I haven't seen any grey Antarctic orca there either, but who knows what the next trip will bring? On the other hand, I have seen two orca in PNG with grey under-flukes – an area on orca that is normally white (even on the grey orca who live down in Antarctica) – so who knows what is going on? As for the orca feeding on sharks in PNG and New Zealand – I don't know who developed the idea first, and who taught whom, or even if the populations never mix and the practice developed independently, but hopefully over the years I will get a better idea of what is happening. I intend to return to PNG to try to obtain some biopsy samples and hydrophone recordings of the local orca. Perhaps these will give us some ideas about whether the animals are indeed related or if they ever share the same waters.

Back in New Zealand I'm still working on trying to get hydrophone recordings of the local orca. I've been lobbying with Terry Hardie to get funding to set up a microwave link from a hydrophone permanently set up at some offshore islands, and we hope this will also give us an early-warning system should orca turn up. John Ford has a similar set-up in the waters off British Columbia so he can listen to any number of different hydrophones along the coast. I dream of adding a voice-recognition software program that notifies me via cellphone when orca calls are heard around New Zealand.

As for some of the other characters you have read about in this book, well, Bill Rossiter from Cetacean Society International is still strongly supporting cetacean research projects. He recently brought some researchers from South America to a large marine mammal conference to give them the chance to network and learn about other research projects that are going on. He and I are often in touch and swapping ideas and research contacts. Jo Berghan had a boy called Callum whose first word, and I kid you not, was 'orca'. Terry Hardie now lives in the US and comes over to help out whenever he can; he still hosts my website and provides assistance for all my computer issues. Tanya Jones has had two girls and she and I have been talking about setting up some educational programmes around New Zealand. Danny Mayson is now married and also has two girls; at present she is living in Australia but intends to come back here to work part-time with orca. Steve Whitehouse is living somewhere in Australia, but I have lost touch with him, while John Ford continues to work with orca in the Pacific Northwest and we hope to work together again, with Terry, to set up remote hydrophones to record those elusive calls. John Craig is still with the University of Auckland and has several other students working with cetaceans these days. Tom and Greta Newhof still support my research through Adopt an Orca USA. My Dad and Chrissy still live in a house overlooking the ocean and call me whenever they see the orca.

And as for the New Zealand orca, they are doing just fine. By the time you read this, some might have died, but others will have been born. I still can't say if the New Zealand population of orca is stable or not, but in reality that will take many, many more years of research to determine. However, as I write this, I can tell you that A1 hadn't been seen for over a year and I was becoming very concerned about her. I thought she must have died, but only a few months ago someone sent me a photo of her. The old girl might have a few more years in her yet – and considering she is the same age as me, I sure hope so. I still don't know if Rinie is a boy or a girl, and probably won't know until he/she is about sixteen years old, unless I get a biopsy sample and we can tell genetically, or I get a good look at the genital area and can sex him/her. Digit is still making close approaches to boats and has only ever been photographed off the north of the North Island, substantiating my concept of the three sub-populations. The last time I saw Miracle she let me rub her down like you would a horse. Meanwhile, Ben is still doing well with his damaged fin – so well in fact that he is now being photographed off the South Island. He had never been seen there before he got hit, so now I need to ask if this is something new, or is it because now that he has such an easily identifiable fin I am getting more photographs of him? It is hard to say, considering he was always a North Island boy and has now been seen off both islands, and I'm beginning to wonder if this is a trend which many of the orca will emulate or if it is just a one-off thing. However, I do have to take into account that Ben was known to mix with Rocky and Spike, both adult male orca who had been to the South Island before. Perhaps they took Ben on a little reconnaissance trip with them? I still don't know what it is that drives some orca to travel further than others and perhaps I never will, but that is what makes this field of study so interesting – to uncover an answer or part of an answer and then realise that this only generates more questions.

As for me, well, I will keep on doing orca research for as long as I can breathe. My passion for these animals is incredibly strong – I guess you could even say that passion has become an obsession. But that's okay, I can live with being obsessed about something as amazing as orca. I intend to keep getting out in the field as often as funding will allow, and eventually hope to build a research centre near Tutukaka in the north of New Zealand, a place where I can have research assistants and fellow researchers come and learn about orca. The orca living around my home patch will always be my primary research focus, but I want to put what I find here into the context of other research projects and other findings I make about orca in different locations.

I intend to keep going to Antarctica to collect more data, and who knows, I might eventually find one of the New Zealand orca at the ice or get a match from the grey Antarctic orca I have seen off New Zealand down in Antarctic waters. Some people think I am crazy to even look, but if we don't look then we'll never find a match. Imagine if we did, how many questions that would pose? And can you envisage how much it would rattle some scientific cages to get a positive match between Antarctica, New Zealand and Papua New Guinea? It may just turn out to be a wild-goose chase, but I'm still keen to look.

In the interim, my next big mission is to get a relisting of the New Zealand orca for the IUCN Red Data List and to propose some appropriate methods to use for rescuing stranded orca. Perhaps these will come in handy for other orca researchers and rescue teams around the world. Should you want to find out more scientific data about orca, or just read more about my research, you can look on the web at www.orcaresearch.org. From this site you can also send in orca sighting reports, as well as adopt an orca or make a donation to support further orca research. I will also be posting notices about any new projects I'm working on and, who knows, maybe even let you know if I'm going to be writing a second book.

# APPENDIX ONE
## Revered and Feared: Orca through the Ages

It is hard to believe, given New Zealand's unique maritime history, and the fact we are surrounded by ocean on all sides, that we have no myths or legends about orca. The Maori arrived here from Polynesia in great ocean-going canoes and have numerous stories about other whale species – including dolphins, sperm whales and southern right whales – but none that pertain to orca.

However, there are other indigenous peoples from around the world with very strong links to orca. Some traditional societies such as the American Indian tribes of the Tlingit, Haida, Tsimshian, Kwakiutl and Coast Salish people revered the orca and regarded them as the spiritual lords of the sea. It was their belief that an orca could drag a boatload of fishermen to the ocean floor and, once there, the people would be transformed into orca themselves. Orca appearing in front of villages were believed to be drowned relatives returning to communicate with their loved ones. The stylised image of the orca, with its high dorsal fin and many-toothed mouth, featured on totem poles as well as numerous everyday items.

Inuit believed that orca could transform themselves into wolves and vice versa, and both species were respected and revered. One myth has the sound of the orca calling as the sound of the earth breathing, and another tells of a huge quake which split the earth and formed the mainland and an island. A tribe which had been living there was also split apart. The people were afraid to go near the water, let alone cross the channel between them, because there was a dreaded black-and-white monster living there. They called it 'Namu', meaning whirlwind. One day a princess from the mainland wanted to try and see her mother who was on the island. So she went down to the ocean and spoke to the monster who offered to give her a ride across on his back. It turned out that the monster was not ferocious at all, just a lonely orca, and all it took to break the spell of fear was someone with the courage to confront him.

In more recent times, however, orca have acquired a far more formidable character, particularly in the Pacific Northwest, where among non-native people the name 'killer whale' is most commonly used to refer to them. In New Zealand the most common name is orca, and although the animals still enjoy a fearsome reputation here, theirs is nothing that rivals the northern hemisphere disposition. This mythical status was perpetuated in many ways. Some of the first field guides to describe animals of the sea often used emotive language which set a tone more of fear than respect. For instance, one such book from the late nineteenth century describing the American whaling industry and marine mammals – *The Marine Mammals of the Northwestern Coast of North America* by C. M. Scammon – said that the orca had 'murderous jaws' and 'in whatever quarter of the world the Orcas are found, they seem always intent upon seeking something to destroy or devour'.

Published in 1912, *Life in the Deep Sea* by F. M. and L. T. Duncan portrayed the orca as an animal with evil ways by describing its behaviour as follows: 'When not engaged in their favourite occupation of killing every creature they can find in the sea, the Killer-Whales sport and jump and tumble and roll in the water, cutting all sorts of capers as if they were rejoicing over their wicked ways.' Apsley Cherry-Garrard, who travelled to Antarctica with Scott during his race to the South Pole, related in his 1922 book, *The Worst Journey in the World*, how the orca (then called *Orca gladiator*) 'were undisguisedly interested in us and the ponies, and we felt that if we once got into the water our ends would be swift and bloody'.

By the late 1950s these kinds of opinions had become well and truly engrained in people's minds. For example, in a book entitled *Dangerous Marine Animals* a section on 'killer whales' stated: 'They are fast swimmers and will attack anything that swims. They have been known to come up under ice floes and to knock seals and people into the water. If killer whales are spotted, the diver should get out of the water immediately.' A 1960s publication called *Man is the Prey* states the case against orca even more explicitly: 'The title of the biggest confirmed man-eater on earth must surely go to this thirty-foot-long member of the dolphin family, the black-and-white killer whale. It could, if it ever had the opportunity, eat a dozen men in one session. There is no animal that swims in the sea or accidentally falls into the sea that will not make a meal for a killer whale.' The author goes on to claim that there are 'numerous stories of killer whales attacking and bumping even quite large boats and no one will ever know how many shipwreck survivors during the war were upset from liferafts and eaten'.

Closer to the present day, a *National Geographic* article from the 1980s contains the extravagant quote: 'The only way to survive an encounter with a killer whale is reincarnation.' Even the United States Navy succumbed to unjustified hysterics about these magnificent animals, when they wrote bluntly in their *Antarctic Sailing Directions* manual: 'Will attack human beings at every opportunity.'

It is hardly surprising then that Hollywood followed suit in the 1977 movie *Orca* in which an adult male orca seeks revenge on humans for the killing of its pregnant mate, and that even today the general public still believe the term 'killer whale' is earned in reference to the eating of people. *Orca*'s byline read: 'The killer whale is one of the most intelligent creatures in the universe. Incredibly, he is the only animal other than man who kills for revenge. He has one mate, and if she is harmed by man, he will hunt down that person with a relentless, terrible vengeance – across seas, across time, across all obstacles.'

Nevertheless, there has been a gradual shift in perceptions of orca over time, and in the mid-1960s Ted Griffith, a marine biologist who had captured an orca which he called Namu (after the legend), began swimming with the animal and discovered that it wasn't going to attack him and that it actively sought out his company instead. Ted grew very fond of the orca and became an adviser to a movie starring Namu, called, ironically, *Namu the Killer Whale*. In the film the orca was released back into the wild, but Namu the real orca died after only a short time in captivity. Unfortunately, his capture initiated the start of the full-scale captive orca industry and since then at least 135 orca have been kept in captivity. It is a sad state of affairs when we consider that many don't survive more than a few years and that they die from a variety of non-natural causes (by August 2004, 113 of those captured orca were dead). The plight of cetaceans in captivity, and in particular orca, warrants some further comment here.

In the contemporary western world we hold high expectations for the way in which other animals should be kept in captivity, yet it continues to amaze me that we allow places such as 'marine parks' – no matter what humanitarian-sounding names they go under – to persist. The animals in these facilities are kept under extremely primitive and cruel conditions. For example, they live in a tank which is nothing more than a blue box, albeit some are a box with curved walls. The walls have no features, no murals painted on them, nothing to distinguish one side from the other, no view to look at other than just a pale blue paint-job. In fact, the paint-job is often so pale that the light reflecting off it appears to be painful to the animals' eyes and the cetaceans trapped in these blue boxes swim around with their eyes shut, or at the very least squinting. Furthermore, these flat and curved walls create acoustical echo-chambers in which sounds emitted by the animals are bounced repeatedly off the walls. The tanks are typically situated very close to the pump rooms, so the animals can hear and feel the vibrations of the pumps running twenty-four hours a day, seven days a week – a constant rumbling in the background which we could liken to living next to a motorway.

As if all that wasn't bad enough, the tanks are so shallow as to prevent any real diving – and given that we know orca can dive to at least 160 metres, and regularly dive to over 100 metres, a tank that is typically five to ten metres deep is not sufficient to provide anywhere near a normal environment. Orca can easily stay underwater for ten minutes, and of course they *could* do that in a tank, but what's the point? In this environment they usually don't stay under longer than two minutes and it doesn't take them very long to lose their fitness and so be no longer able to maintain such diving feats. In the wild they might swim over 100 kilometres in just one day – in a tank it is unlikely they will swim more than 100 laps, and again, through lack of exercise, lose their physical conditioning.

In their natural environment wild orca spend less than ten per cent of their lives at the surface, so while underwater they avoid the harmful effects of UV and other rays from the sun. While diving they have constant water pressure exerted on their bodies which probably helps to maintain their rigid dorsal fins, which in captivity droop irreversibly. Overall, the decline of their health in captivity is most likely linked to a number of factors, some of which are physical, such as fitness; the type of food the animals are fed (frozen fish); and the chemicals used to treat the water in which they swim (as there is urine and faeces in the tank water, these chemicals often have to be abundant and strong to prevent any number of diseases or other health problems).

There is also the matter of the mental health of any cetacean in captivity to be considered, including removal of the individual from its social group. We now know that for some groups of orca the only way they enter a group is through birth and the only way they leave it is through death – unless the animal is caught for display in one of our 'marine parks'. To remove an orca from the context of its group is nothing short of barbarity. Often they are placed in tanks by themselves – so they have no acoustical, visual or physical stimulation from other orca. Or they might be put in tanks with other cetacean species, such as bottlenose dolphins, which in some cases would be their natural prey. In rare instances they are placed in tanks with other orca, who may in fact not 'speak' the same dialect as them, and can also be openly hostile to newcomers.

The three main arguments for the continuation of keeping cetaceans in captivity revolve around 'education', 'conservation' and 'ambassadorship'. In the first case, the pro-captivity people – generally the undereducated public and the profit-earning owners of the parks – maintain that we are learning about these animals by keeping them in captivity, and that by displaying them to the general public we can educate them about what wonderful creatures they are. They also argue that valuable research on orca is conducted in captivity.

To some degree, I guess you could say that the proponents of 'education' are right. To the average person, seeing an orca in captivity is a truly amazing sight. A huge black-and-white creature, totally dominated by their diminutive trainer (usually a good-looking girl), must be a visual treat. But to my mind, if you are going to teach people about ducks, you don't show them a Donald Duck cartoon. Orca in captivity are not a true representation of orca in the wild.

They are a caricature, a comic strip – only it's far from funny – of the real thing. It isn't educational to teach people that it's okay to mistreat animals and keep them in inappropriate cages without social, mental or physical stimulation. Showing people being 'rocketed' through the air on the end of an orca, or surfing an orca around a tank isn't educational. Entertaining, no doubt. Money-making, for sure. Amazing to watch, absolutely. Educational, definitely not. Yes, some institutions make an effort at educating their audiences about the animals, but this is often out-of-date, extremely 'filtered' information which is superfluous to the ritz and glitz of the show.

Then there is the 'conservation' aspect. Again, I might have to concede that there are possibly some conservational benefits which *might* come out of keeping orca in captivity. For instance, one marine park raised money to support conservation of orca in the wild, including an educational programme, but this is not the industry standard. That particular park also supported research on wild orca – and continues to do so today, even though they no longer hold orca in their facility. Their wild-based research was, and continues to be, very valid science which is published internationally and is available to other scientists and the public alike. Other facilities state that they are conducting research, and no doubt they are, but this is all done on the animals in their tanks. And again, to my mind, if you want to study, say, human behaviour, you don't put someone in a telephone box and expect to obtain valid results. The same objection applies to placing an orca in a tiny tank and monitoring its behaviour – or any other aspect of its life for that matter. Also, although these kinds of institutions claim they are conducting research, very little of it actually makes it out into the public forum. There is good commercial reason for this of course, because if you have found a way to keep your animals alive, then you don't want the competition to find out about it, and if you haven't found a way and your animals are dying then you also don't want anyone to find out.

Currently the United States holds the most orca in captivity – twenty-three animals – followed by Japan who have twelve and France who have six. The newest area for capturing wild orca is Russia, particularly the Kamchatka region, and most of these animals are destined for the Japanese captive market. Rumour has it that a female orca sells on the open market for over US$1 million dollars. That's a lot of money, but not a bad investment when you consider that a ticket to enter a typical marine park in the US costs around US$50 per person (if you want to eat lunch next to an orca tank it *only* costs $30 extra), and one park has had over 100 million visitors since it opened in the mid-sixties. Another way some parks raise income from their captive orca is to offer meetings 'up close and personal', with some of the parks offering a 'day as a trainer' for *only* US$400 (but hey, it includes a free T-shirt). And that's not to mention all the profits made from the sale of orca merchandise through the park gift shops.

Some apologists for the marine parks argue that captive breeding is essential to preserve the species. However, until just recently, no population of orca had been designated as under threat, and the success of orca breeding in captivity has been dismal. Advocates for the captivity industry say that breeding orca are happy orca; however my response to them would be that women in concentration and refugee camps have children, and that doesn't mean that they were or are happy. Of the sixty-four known orca pregnancies in captivity worldwide since 1968, only twenty-seven captive-born calves (less than half) have survived. Furthermore, by putting an orca in captivity you have guaranteed that it will die much sooner than it would in the wild. Given that in the wild the estimated life span for orca is approximately fifty years for females and twenty-nine years for males, you would expect at least half that extent for them in captivity (and based on some other species of animals in captivity, perhaps even an extension of their life span), but alas, that isn't the case. Orca in captivity rarely live longer than five years past their capture date, and to date none has lived longer than twenty-eight years (and that one individual spent the last five years of his life in a semi-wild situation and then died).

That individual was a special case. He was known to many people around the world as 'Willy', but his real name was 'Keiko'. He was the star of the *Free Willy* movies where a young boy befriends an orca in captivity and helps him return to the wild. At the end of the movie Willy returned to his family in the wild (just like in the movie *Namu*), but in reality he was stuck in a tiny tank in Mexico – an unwilling 'ambassador' from his natural domain. Many people who saw the movie (in particular kids) were not impressed when they found out that Keiko didn't live the fairy-tale ending of the film script. So substantial amounts of money were raised to try to release him back into the wild. Keiko's story illustrated a growing public disillusionment with the captive cetacean industry and a significant change in general attitudes towards orca – the byline for this movie was: 'How far would you go for a friend?'

And speaking of 'ambassadors', human ambassadors usually choose their vocation and are not normally separated from their families when they take up their post. Captive orca are nothing more than unwilling prisoners, permanently ripped from their social groups. What human ambassadors are kept locked up in a cramped blue box, fed only when they perform tricks and *never* allowed to return home? If we truly regarded captive orca as ambassadors, we would treat them with more respect.

How people respond to the natural world influences their behaviour, and within the last decade the human perception of apex predators has changed. For instance, wolves and mountain lions were often presented as the embodiment of all evil but have now become powerful and popular symbols of conservation groups. It is also becoming apparent that another predator whose image has undergone a metamorphosis is the orca.

On a worldwide scale, cetaceans in general have a high conservation profile and are often used as 'poster animals' for the conservation movement. There has been strong public pressure to protect them from threats such as whaling. In New Zealand, since the early 1900s, orca were taken as by-catch for oil and for so-called 'scientific' purposes until at least 1979. They have been hunted on a commercial scale in Antarctica where more than 900 were taken in 1980 alone; so considering that New Zealand orca may be travelling to Antarctic waters they might well be under threat.

Given all the negative aspects of keeping cetaceans in captivity, the general trend to want to protect these animals, and the growing popularity of eco-tourism offering people the opportunity to see things in 'nature' and in the 'wild', there is ever-increasing interest in whale watching. Orca are generally considered one of the most spectacular of all cetaceans encountered on whale-watching trips, with people strongly attracted to their vivid black-and-white colouring and their 'top predator' aura. Still, this change in attitude has taken time, and even today many people still fear being tipped out of boats or being attacked while in the water. Education about these animals in the wild will be the key to transforming such misconceptions.

A whole science has sprung up surrounding how to study these creatures in their natural environment, and keeping the public informed about orca is a key role of the scientists working with them. Yet even after studying orca for over thirty years we still know surprisingly little about these awe-inspiring cetaceans which have become the focus of my life's work.

# APPENDIX TWO
## Extracts from my PhD Thesis

The following material is some extracts from my PhD chapter on Associations. In layperson's terms, this is about which orca were hanging out with which, how often, and how that information fits with the socially complex patterns of an animal such as the orca. To read about the context of these results, see Chapter Ten. This section explains the use of an Association Index, which is a numerical way of referring to the number of times any two orca are seen together. However, there are problems which can arise from trying to understand how often two animals are seen together, and this is also explained below. References to other scientific publications are listed at the end of the appendix.

The Association Index chosen followed Cairns and Schwager (1987) who suggested a formula they refer to as the Half-Weight Index, which would be the least biased for studies where animals are more likely to be recorded separately than when together. This bias is likely to occur in photographic studies (which are typically conducted for cetaceans), where the number of mutual sightings is likely to be underestimated. There are two reasons why this might happen: first, before two individuals can be scored as sighted together, both must be seen and photographed; secondly, when two individuals are separate this can be recorded if either of the two individuals is photographed, whereas only one individual can provide association data when they are together (Cairns and Schwager 1987).

Therefore, the Half-Weight Index is commonly used in studies of cetacean associations, e.g., spinner dolphins (*Stenella longirostris*) (Östman 1994), bottlenose dolphins (*Tursiops truncatus*) (Bräger et al. 1994, Conner et al. 1992, Schneider 1999), Hector's dolphin (*Cephalorhynchus hectori*) (Bejder and Dawson 1997) and orca (Heimlich-Boran 1987).

The Half-Weight Index formula is:

$$\frac{x}{x + \frac{1}{2}(yA + yB)}$$

where $x$ is the number of sightings that included both animal A and animal B in the same group,

$yA$ is the number of sightings that included animal A, but not animal B, and

$yB$ is the number of sightings that included animal B, but not animal A.

The higher the Association Index (i.e., the closer the number is to 1), the more time the animals spend together. A value of zero would indicate that two animals were never seen together. When using an Association Index that is limited to analysis of dyads (pairs) by default, the test of association is between two animals. However, in a group of, for instance, three orca, each animal present is considered to have two associations, but each would be calculated independently. Consequently any multiple associations (no matter how many orca were present in a group) were all calculated separately, as if they were dyads. In addition, animals were recorded as being together only once for any particular sighting event. However, some of these animals were in close association for periods of up to 14 hours, whereas others were observed closely associated for time periods as short as two hours. Hence the Association Index only measures one dimension of social association (in this case, being seen together, but not others, such as for how long).

The highest Association Index value was (0.93), which was calculated for four dyads. The next highest Indices (0.84) and (0.83) were calculated for one dyad each, and (0.80) was calculated for four dyads. The mean Association Index value was 0.25, the mode was 0.13 and the median was 0.18. However, it should be noted that this included 857 instances where there was no association at all (i.e., an Association Index of 0).

Fifteen orca with the highest Association Indices are presented in Table 1. Fourteen of these orca have interacted with 10 or more orca, who were also seen more than five times. One adult female (NZ1 – 'A1') was seen with 48 other individuals from the total photo-identified population ($n = 117$), of which 21 were also seen on more than five occasions. She associated more often with certain individuals than others, e.g., NZ3 (Olav) (0.93), NZ48 (0.67), NZ49 (0.50), NZ99 (0.37), compared to NZ6 (Rocky) (0.25) and NZ27 (Yin) (0.13). Although some individuals had a large number of associates, the Association Index values suggest that close associations are limited to as few as three other animals, e.g., NZ7 (Spike) was seen with NZ8 (Sergeant) (0.53), NZ9 (Flean) (0.63), and NZ13 (Double Notch) (0.67); NZ25 (Prop) was seen with NZ44 (Ragged Top) (0.64), NZ88 (0.86) and NZ90 (0.67); and NZ44 (Ragged Top) was seen with NZ88 (0.75), NZ89 (0.75) and NZ25 (Prop) (0.64).

**Table 1** Individual orca and the number of associates

| Orca Catalogue Number | Orca Name | Total Number of Associates | Number of Associates who were also seen more than five times |
|---|---|---|---|
| NZ1 | Al | 48 | 21 |
| NZ3 | Olav | 43 | 20 |
| NZ4 | Venus | 22 | 11 |
| NZ5 | Sickle Fin | 30 | 11 |
| NZ6 | Rocky | 35 | 12 |
| NZ7 | Spike | 33 | 10 |
| NZ8 | Sergeant | 35 | 20 |
| NZ9 | Flean | 31 | 10 |
| NZ13 | Double Notch | 30 | 13 |
| NZ24 | Rudie | 27 | 15 |
| NZ25 | Prop | 16 | 11 |
| NZ26 | Top Notch | 15 | 10 |
| NZ27 | Yin | 28 | 16 |
| NZ44 | Ragged Top | 13 | 7 |
| NZ101 | Ben | 30 | 20 |

Some associations were long term, e.g., NZ1 (A1 – adult female) had been seen around the New Zealand coastline for 20 years, and had been sighted with NZ3 (Olav – adult male) on 19 occasions during 12 years. NZ15 (Corkscrew – adult male) was first sighted in 1985 with NZ16 (Nicky – adult female) and was resighted with the same female in 1994 (i.e., nine years later).

And here is what the circle-plots looked like. I had to use two, because there were two very clearly different association levels going on in the world of New Zealand orca. The first one shows four distinctive groups of orca. For three of these, most of the lines are thick and there are multiple lines between the animals (meaning that they all hang out together a lot). The fourth group had two orca in it and they had a fairly thin line joining them (Association Index of just under 0.50). These two were a mother-and-calf pair and the calf was only a few years old, so I didn't have a lot of data on them. For this circle-plot all the orca were only seen with orca from their same sub-population. That is, if they were a 'North Island' orca then they were seen with other 'North Island' orca and not with orca who had been seen off both the North and South Islands (by default they couldn't be seen with orca from the 'South Island' population, as geographically the two populations wouldn't meet). These distributions were indicated by a line placed above or below the orca code (see Chapter Ten for a fuller explanation).

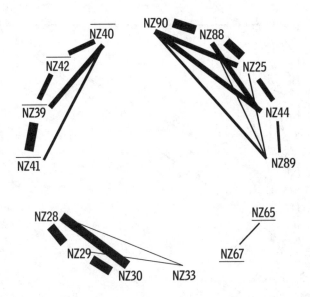

For the second circle-plot things got a little more complicated. I found that there were so many lines going from some orca (such as Ben, NZ101, on the middle left of the circle-plot) that it was like a spider's web. But this was interesting in itself. It showed that there weren't only the strong bonds between Ben and NZ95 (another young male) and NZ96 (an adult female, perhaps Ben's mother?) but also that there were social bonds between Ben and Digit (NZ50)

and Dian (NZ51). Overall there did appear to be some grouping – and I arranged the animals that had the strongest bonds next to each other around the edge of the circle, but it was not as clear-cut as the first circle-plot. And in this example I could see that orca who had only been seen off the North Island (like Ben) were seen with orca who had been seen off both the North and South Islands (like NZ7 – Spike).

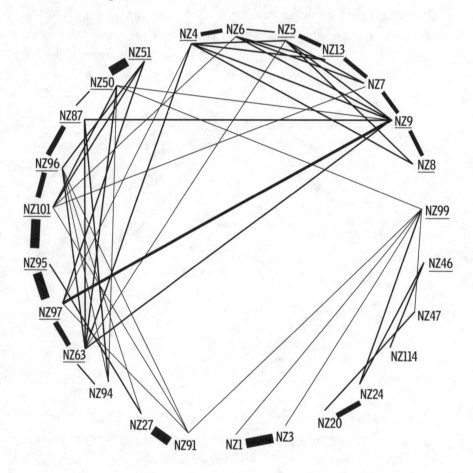

All in all, the associations, their indices, the circle-plots, and what I was witnessing out in the real world were complicated and intriguing – and all the more reason to continue studying these fascinating creatures who have become my friends and the centre of my life.

Bejder, L. and Dawson, S. (1997). 'Ecology and behaviour of Hector's dolphin and tourism in Porpoise Bay', unpublished interim report to the World Wide Fund for Nature, New Zealand, Department of Marine Science, University of Otago, PO Box 56, Dunedin, New Zealand.

Bräger, S., Würsig, B., Acevedo, A. and Henningsen, T. (1994). 'Association patterns of bottlenose dolphins (*Tursiops truncatus*) in Galverston Bay, Texas', *Journal of Mammalogy* 75, 431–37.

Cairns, S. J. and Schwager, S. J. (1987). 'A comparison of association indices', *Animal Behaviour* 35, 1454–69.

Conner, R. C., Smolker, R. A. and Richards, A. F. (1992). 'Two levels of alliance formation among male bottlenose dolphins (*Tursiops* sp.)', *Proceedings of the National Academy of Science, USA*, 89, 987–90.

Heimlich-Boran, J. R. (1987). 'Habitat use patterns and behavioral ecology of killer whales (*Orcinus orca*) in the Pacific Northwest', Thesis, San Jose State University, California, USA.

Östman, J. S. O. (1994). 'Social organization and social behaviour of Hawai'ian spinner dolphins (*Stenella longirostris*)', PhD Dissertation, University of California, Santa Cruz, USA.

Schneider, K. (1999). 'Behaviour and ecology of bottlenose dolphins in Doubtful Sound, Fiordland, New Zealand', PhD Dissertation, University of Otago, Dunedin, New Zealand.

## Scientific articles by Dr Ingrid N. Visser

Berghan, J. & Visser, I. N. (2000). 'Vertebral column malformations in New Zealand delphinids with a review of cases world-wide', *Aquatic Mammals* 26, 17–25.

Berghan, J. & Visser, I. N. (2001). 'Antarctic Killer Whale Identification

Catalogue', 14th biennial conference on the biology of marine mammals, 28 November–3 December, 2001, Vancouver, Canada, p. 22.

Constantine, R., Visser, I., Buurman, D., Buurman, R. & McFadden, B. (1998). 'Killer whale (*Orcinus orca*) predation on dusky dolphins (*Lagenorhynchus obscurus*) in Kaikoura, New Zealand', *Marine Mammal Science* 14, 324–30.

Duignan, P. J., Hunter, J. E. B., Visser, I. N., Jones, G. W. & Nutman, A. (2000). 'Stingray spines: A potential cause of killer whale mortality in New Zealand', *Aquatic Mammals* 26, 143–47.

Visser, I. N. (1998). 'Prolific body scars and collapsing dorsal fins on killer whales in New Zealand waters', *Aquatic Mammals* 24, 71–81.

Visser, I. N. (1999). 'Antarctic orca in New Zealand waters?', *New Zealand Journal of Marine and Freshwater Research* 33, 515–20.

Visser, I. N. (1999). 'Benthic foraging on stingrays by killer whales (*Orcinus orca*) in New Zealand waters', *Marine Mammal Science* 15, 220–27.

Visser, I. N. (1999). 'Propeller scars and known migration of two orca (*Orcinus orca*) in New Zealand waters', *New Zealand Journal of Marine and Freshwater Research* 33, 635–42.

Visser, I. N. (1999). 'A summary of interactions between orca (*Orcinus orca*) and other cetaceans in New Zealand waters', *New Zealand Journal of Natural Science* 24, 101–12.

Visser, I. N. (2000). 'Killer whale (*Orcinus orca*) interactions with longlines fisheries in New Zealand waters', *Aquatic Mammals* 26, 241–52.

Visser, I. N. (2000). 'Orca (*Orcinus orca*) in New Zealand waters', PhD Dissertation, University of Auckland, Auckland, New Zealand.

Visser, I. N. (2005). 'First observations of feeding on thresher (*Alopias vulpinus*) and hammerhead (*Sphyrna zygaena*) sharks by killer whales (*Orcinus orca*) which specialise on elasmobranchs as prey', *Aquatic Mammals* 31, 83–8.

Visser, I. N. & Bonaccorso, F. J. (2003). 'New observations and a review of killer whale (*Orcinus orca*) sightings in Papua New Guinea waters', *Aquatic Mammals* 29, 150–72.

Visser, I. N., Fertl, D., Berghan, J. & van Meurs, R. (2000). 'Killer whale (*Orcinus orca*) predation on a shortfin mako shark (*Isurus oxyrinchus*), in New Zealand waters', *Aquatic Mammals* 26, 229–31.

Visser, I. N., Fertl, D. & Pusser, L. T. (2004). 'Melanistic southern right-whale dolphins (*Lissodelphis peronii*) off Kaikoura, New Zealand, with records of other anomalously all-black cetaceans', *New Zealand Journal of Marine and Freshwater Research* 38, 833–36.

Visser, I. N. & Fertl, D. C. (2000). 'Stranding, resighting and boat strike of a killer whale (*Orcinus orca*) off New Zealand', *Aquatic Mammals* 26, 232–40.

Visser, I. N. & Mäkeläinen, P. (2000). 'Variation in eye patch shape of killer whales (*Orcinus orca*) in New Zealand waters', *Marine Mammal Science* 16, 459–69.

# GLOSSARY
## Orca Terminology

The following are terms that are used in this book and in orca research.

### Antarctic orca
Orca which are found south of 60°, in Antarctic waters. There are three types: **Type A** which is black and white and looks like the 'typical' orca; **Type B**, which is grey and white and has a big **eye patch** and a **dorsal cape**; and **Type C**, which is also grey and white, has a **dorsal cape** and a small **eye patch**

### Association Index
A number calculated from sightings of individual orca, used to represent how much time they spend together

### breach
A leap out of the water, where at least two-thirds of the body is out of the water (also see **half-breach**)

### calf
An orca which is less than half the size of the mother – typically less than two years old

### cetacean
A whale, dolphin, or porpoise

*dialect*
A unique set of orca calls which are made by a group of whales, and individuals within that group

*dorsal cape*
An area of pigmentation, found only on Antarctic orca, which runs from the **eye patch** to the **saddle patch**. The dorsal cape is generally darker than the sides of the orca

*dorsal fin*
The fin on the back of a **cetacean**

*elasmobranch*
A collective term to describe sharks and rays

*eye patch*
The white, typically oval-shaped patch on the side of the face of an orca. The eye is found just below and forward of the eye patch

*flukes*
The horizontal part of the tail of an orca

*food-share*
Two or more orca sharing food – typically a prey item which is too large to be eaten in one bite by an orca

*half-breach*
A leap where just the front half of the body comes out of the water

*hydrophone*
An underwater microphone used to listen and record underwater sounds. It can not play sounds back (this would require underwater speakers)

*juvenile*
An immature orca, of either sex, which is larger than a calf, and smaller than a sub-adult orca

## *killer whale*
An alternative name for orca. Usually used in the Pacific Northwest (i.e., off the West Coast of North America)

## *matriarch*
An older female orca, typically the head of the group

## *Offshore orca*
A population of orca living off the Pacific Northwest about which little is known. They typically travel in large groups and are thought to eat fish

## *Orcinus glacialis*
Latin (or scientific) name for a form of orca found in Antarctic waters which may be a separate species from **O. orca**. First described by Russian scientists, they are most likely to correspond to the **Type C** orca, which are typically found in larger groups than **Type A** or **Type B**, and are thought to feed primarily on fish, but may occasionally take penguins. Typically found around dense pack ice or fast ice, hence the name, meaning glacier or ice. See **Antarctic Orca**

## *Orcinus nanus*
Latin name for a 'dwarf' form of orca found in Antarctic waters which may be a separate species from **O. orca**. First described by Russian scientists, they are most likely to correspond to the **Type B** orca. O. *nanus* are thought to feed primarily on pinnipeds (seals etc.), but may occasionally take penguins. Typically found around loose sea ice, they often **spy hop** to investigate the top of ice floes for potential prey. See **Antarctic orca**

## *Orcinus orca*
Latin name for orca. Currently thought to be the only species of orca found throughout the world, but evidence is mounting that there may be other species, particularly in Antarctic waters. See **Antarctic orca**, **O. glacialis** and **O. nanus**

*pectoral fin*
The side fins of a cetacean (their 'arms')

*pod*
A group of orca which usually travel together for more than fifty per cent of the time

*Resident orca*
Populations of orca living off the Pacific Northwest, which typically travel in large groups, eat fish (usually salmon) and have very stable social groups, which are matriarchal (see **matriarch**)

*saddle patch*
An area of grey pigmentation behind the dorsal fin of orca. It is variable in shape and size

*sub-adult male*
A male which has begun to grow his large dorsal fin, but is not yet fully grown (a 'teenager'). He could also be called a 'sprouter' as his fin has begun to sprout

*spy hop*
When an orca swims upwards and raises its head above the water. Typically in a spy hop the pectoral fins would also be visible

*Transient orca*
Populations of orca living off the Pacific Northwest, which travel in small groups (sometimes as few as one individual). They typically eat marine mammals, such as seals and other cetaceans, but also birds. They have a less regimented social structure compared with **Resident** orca

*Type A, Type B, and Type C orca*
See **Antarctic orca**